Sprinkler Irrigation

RK Sivanappan

Oxford & IBH Publishing Co. Pvt. Ltd.
New Delhi
(A Unit of CBS Publishers & Distributors Pvt Ltd *)*

CBS

CBS Publishers & Distributors Pvt Ltd

New Delhi • Bengaluru • Chennai • Kochi • Kolkata • Mumbai
Bhopal • Bhubaneswar • Hyderabad • Jharkhand • Nagpur
Patna • Pune• Uttarakhand • Dhaka (Bangladesh)

Sprinkler Irrigation

ISBN-13: 978-81-204-0232-4
ISBN-10: 81-204-0232-4

OXFORD & IBH
New Delhi
(A Unit of CBS Publishers & Distributors Pvt Ltd)

Published by Satish Kumar Jain and produced by Varun Jain for
CBS Publishers & Distributors Pvt Ltd
4819/XI Prahlad Street, 24 Ansari Road, Daryaganj, New Delhi 110 002, India.
Ph: 23289259, 23266861, 23266867   Fax: 011-23243014   Website: www.cbspd.com
e-mail: delhi@cbspd.com; cbspubs@airtelmail.in.
Corporate Office: 204 FIE, Industrial Area, Patparganj, Delhi 110 092
Ph: 4934 4934   Fax: 4934 4935   e-mail:publishing@cbspd.com;
                                                                                    publicity@cbspd.com

Branches

- **Bengaluru:** Seema House 2975, 17th Cross, K.R. Road, Banasankari 2nd Stage, Bengaluru 560 070, Karnataka
  Ph: +91-80-26771678/79   Fax: +91-80-26771680   e-mail: bangalore@cbspd.com
- **Chennai:** 7, Subbaraya Street, Shenoy Nagar, Chennai 600 030, Tamil Nadu
  Ph: +91-44-26260666, 26208620   Fax: +91-44-42032115   e-mail: chennai@cbspd.com
- **Kochi:** 42/1325, 1326, Power House Road, Opp KSEB Power House, Ernakulam 682 018, Kochi, Kerala
  Ph: +91-484-4059061-65   Fax: +91-484-4059065   e-mail: kochi@cbspd.com
- **Kolkata:** No. 6/B, Ground Floor, Rameswar Shaw Road, Kolkata-700014 (West Bengal), India
  Ph: +91-33-2289-1126, 2289-1127, 2289-1128   e-mail: kolkata@cbspd.com
- **Mumbai:** 83-C, Dr E Moses Road, Worli, Mumbai-400018, Maharashtra
  Ph: +91-22-24902340/41   Fax: +91-22-24902342   e-mail: mumbai@cbspd.com

Representatives

• Bhopal	0-8319310552	• Bhubaneswar	0-9911037372	• Hyderabad	0-9885175004
• Jharkhand	0-9811541605	• Nagpur	0-9021734563	• Patna	0-9334159340
• Pune	0-9623451994	• Uttarakhand	0-9716462459		
• Dhaka (Bangladesh)	01912-003485				

Printed at Chaman Enterprises, Daryaganj, New Delhi, India

Foreword

Land and water are major natural resources essential to produce food and fibre to the world's ever-increasing population. While the land resources remain constant, the average annual precipitation, in spite of regional variations, also remain nearly constant. The gross cropped area, however, can be increased manifold by increasing the intensity of cultivation. This would necessitate the development and application of the technology to conserve the water resources and increase the efficiency of their utilization. The intensity of cropping in India is presently around 120 per cent. This could be increased to 200 per cent or more by the application of improved agricultural technology.

Water is often the limiting factor to achieve the production potential of a region. The problem is acute in the arid and semi-arid areas with comparatively low rainfall and frequent drought periods. In a monsoon dominated climatic region like the Indian sub-continent, even in regions of heavy rainfall prolonged drought conditions often affect adversely even plantation and orchard crops. Adoption of water-saving irrigation methods like sprinkler and drip irrigation provide the means of increasing the efficiency of irrigation. It enables considerable additional land to be brought under irrigation with a given quantity of water. The efficiency of the system, however, would require its design and operation not only in terms of performance requirements but also in terms of specific crop needs and cropping systems.

Prof. R.K. Sivanappan is an authority in the area of sprinkler and drip irrigation, having spent over three decades of research and development in the area. The present book presents sufficient details of design and application of sprinkler irrigation. It also presents the need and scope for research and development in this vital field. The work includes the selection of the type of sprinkler system for specific locations, including the selection of suitable pumping plant and power requirement. Information on when and how much to irrigate, including inter-related soil-water and plant relationships are presented. Details of mainten- ance of the system are given. The book is written in simple lan- guage and is supported with valuable information presented in tables, sketches and graphs to make easy the understanding of the subject.

I congratulate Prof. R.K. Sivanappan for this valuable contri- bution. I am confident that the book will be of great benefit to all those working in the field of irrigation water management, especially the design and application of the sprinkler system.

A.M. MICHAEL
B.Sc. Agril. Engg.; M. Tech.;
Ph.D.; F.I.E. (India); F.I.S.A.E.
Director
Indian Agricultural Research Institute
New Delhi 110012

Preface

This book is intended as a professional text for students of Agriculture and Engineering faculties and reference in allied disciplines. In addition, it will be a valuable reference to professional engineers and agricultural scientists working in the area of irrigation and water management. This book also will be useful for the bankers, estate owners, farmers and others who deal with sprinkler irrigation.

Sprinkler irrigation was not very popular in India. Regional disparities in water resource availability and the water scarcity frequently encountered due to monsoon failure and unfavourable topographical and soil conditions have focussed the importance of water saving irrigation methods not only in hills for plantation crops, but also in the plains for a variety of crops. The Central and State Governments are keen in introducing on a large scale water saving methods of irrigation. Sprinkler irrigation involves heavy investment in installation and operation. Commercial banks are coming forward to lend money to farmers.

The present work is a comprehensive treatment on the subject, providing the needed technology with special reference to the agro-climatic and socio-economic situations in India.

The details regarding the design and layout, including step-by-step procedures for various crops with model designs are provided. Further, it deals with various other uses of the system, pumping plants, operation and care of the system and when and how much to irrigate. The economic feasibility of the system and research needs are also described. Tables, sketches and photographs are numbered chapterwise. A list of important references is given. The entire book has been written using metric units of weights and measures. Conversion tables for change of units from one system to another are given at the end.

vi *Preface*

The author is deeply indebted to many individuals and organisations for the supply of useful materials for the text and the assistance provided at various levels for the preparation of the manuscript. Special mention is made for the valuable help received from Dr. S.R. Sree Rangaswamy, Director, School of Genetics, TNAU and Dr. S.P. Palaniappan, Prof. & Head of the Department of Agronomy, TNAU for their encouragement and review of the manuscript. My sincere thanks are due to Mr. Chandragiri, Asstt. General Manager and his colleagues of Premier Irrigation Equipment Ltd. for providing many photographs to be incorporated in the book. I am very grateful to Dr. A.M. Michael, Director, IARI, New Delhi for writing a foreword for this book.

Coimbatore   R.K. Sivanappan

Contents

Contents

CHAPTER 1

Introduction

Irrigation is as old as civilisation, but the development or advancement of irrigation is not comparable to any other field. The importance of irrigation in increasing yields from agricultural and horticultural crops for the growing population has been recognised for many years. This has necessitated the rapid expansion of irrigation throughout the world. Water being a limited resource, its efficient use is basic to the survival of the ever increasing population of the world.

Irrigation is the artificial application of water to soil for the purpose of crop production. Irrigation water is supplied when the water is not available through rainfall and the contribution to soil moisture from the ground water is not sufficient for the crop growth. In Europe, in the eastern parts of the U.S.A. and in many other countries, crops are grown without any irrigation since the rainfall to the required amount is distributed throughout the crop period. But in these places, due to climatic changes in the recent years, supplemental irrigation by means of sprinklers is given during the long gaps between rains in order to maintain the crop yield. In countries like India where crops can be grown throughout the year but the rain falls only during the monsoon period for three to four months and that too in an erratic way, supplemental irrigation is a must to grow any crop for any period, except in some locations for a season.

Water is the essence of life on which man depends not only for his direct consumption needs, but also for producing his food and fibre and for various other day to day requirements. This is the reason why early civilisation took place by the sides of perennially flowing rivers such as the Nile, Euphrates, Tigris and the Indus.

The increase in human population coupled with rapid agricultural and industrial expansion has resulted in a sharp rise in demand for water in all places. Though the available quantities may be sufficient to meet the overall requirements, it is unevenly distributed and often found either in inaccessible places or when it is not needed. It is not always the supply of water, but the ability to fully and efficiently utilise the available quantities which restricts economic progress. The primary concern of the Government/individuals/organisations has been to construct dams, channels and wells with a view to augment the quantities of irrigation water available to the farmers. But due attention has not been paid to use the water efficiently by improving management and allocation practices. This has created soil problems which have reduced the productivity of the irrigated land.

Many of the developing countries are experiencing acute water shortage resulting from a fast rate of growth in population, agricultural expansion and industrial development.

The application of outdated irrigation practices has resulted not only in considerable losses of water, but also widespread salinity and water-logging problems and has disturbed the ecosystem adversely. Most irrigation projects have not achieved a substantial increase in agricultural production as was anticipated. Judicious utilisation of the available irrigation water is an essential factor for the productivity of scarce resources. Depending upon the availability of land and water, strategies should be worked out to get the maximum yield per unit land and unit water.

Irrigation history

In India, irrigation has been practiced as an art for about 3000 years now. Historical records bear testimony to the existence of a number of irrigation works in different parts of the country. The character of these works was largely conditioned by the physiographical features of the area in which they are located. In the north, the perennial river Ganges made it relatively easy to divert its flow through inundation channels. In the south, where the rainfall is scanty, the practice of trapping storm water in large tanks and ponds for agricultural purposes is widely adopted.

Under the British rule, irrigation works began with the renovation, improvement and extension of existing works. A number of irrigation projects, namely Upper Ganga canal, Upper Basin Doab canal, the Krishna and the Godavari delta system were completed by 1866, and the other major projects like Sorbind, Lower Ganga, Agra and Mutha canals, the Periyar Irrigation Projects and smaller works like the Pennar River Canals were taken up later. At the end of the nineteenth century the gross irrigated area was about 7.5 million ha by public works.

The first irrigation commission was appointed in 1901, specially to report on irrigation as a means of protection against famine in India. As a result of the Commission's recommendations, the total irrigated area increased to 16 million ha by 1924-25. The progressive increase of the irrigated area in India from 1900 to 1950 is given in Table 1.

Table 1. Progressive development of irrigation in India
from 1900 to 1950

Period year	Net area irrigated from		Total (all sources) in Mha
	Govt. Canal Mha	Wells Mha	
1900-1901	—	—	13.2
1914-15	4.4	4.0	14.5
1919-20	4.1	4.7	15.7
1924-25	4.4	4.7	16.0
1929-30	4.6	4.8	16.2
1934-35	5.0	4.8	17.1
1939-40	5.6	5.2	18.0
1944-45	6.0	5.4	19.0
1949-50	6.4	6.3	19.4

Source: Report of the Irrigation Commission, 1972.

Progress during the Five Year Plans

The net area irrigated in 1950-51 was 22.60 million ha. The Planning Commission gave great importance and priority to develop irrigation for increasing agricultural production. Major projects like the Bhakra Nangal, Damodhar Valley, Hirakud, Nagarjunasagar, Kosi, Chambal, Tungabhadra and Lower Bhavani were taken up. By the end of March 1985, India completed six Five Year Plans and five annual plans. Table 2 gives the area brought under irrigation during the five year plans and the ultimate feasible area.

Table 2. Irrigation potential and area brought under irrigation during five year plans (in Mha)

Plan period	Major & medium	Minor irrigation		Total irrigation
		Ground water	Surface	
1. Preplan (1950-51)	9.70	6.50	6.40	22.60
2. End of 1st plan (1955-56)	12.19	7.63	6.43	26.25
3. End of 2nd plan (1960-61)	14.33	8.30	6.45	29.08
4. End of 3rd plan (1965-66)	16.56	10.52	6.48	33.56
5. End of annual plan (1968-69)	18.10	12.50	6.50	37.10
6. End of 4th plan (1973-74)	20.70	16.50	7.00	44.20
7. End of 5th plan (1978-79)	24.77	19.80	7.50	52.25
8. End of Annual plans	27.22	22.10	8.00	57.50
9. End of 6th plan (targeted 1985)	32.72	29.10	9.50	71.32
10. Ultimate feasible 2000/2025	58.50	40.00	15.00	113.50

Source: Water Resources Development in India, American Embassy, New Delhi, 1980.

In addition to the major and medium projects, minor irrigation works and specially the ground water schemes have also been receiving greater attention. Wells in large numbers were constructed by the farmers with the help of the Government who provided them with finance through land development and nationalised banks to do so. The main advantage of this is that the farmers can regulate the water supply depending upon their needs to irrigate their farms. The state-wise potential area irrigated by different sources of water supply including wells is given in Table 3.

Table 3. State-wise irrigation potential source-wise (in Mha)

Sl. No.	Name of the State	Major and medium	Minor irrigation		Total ultimate irrigation
			Surface	Ground water	
1.	Andhra Pradesh	5.0	2 0	2.2	9.2
2.	Assam	0.97	1.0	0.7	2.67
3.	Bihar	6.50	1.90	4.0	12.40
4.	Gujarat	3.00	0.05	1.4	4.75
5.	Haryana	3.00	0.05	1.4	4.45
6.	Himachal Pradesh	0.05	0.25	0.05	0.35
7.	Jammu and Kashmir	0.25	0.40	0.15	0.80
8.	Karnataka	2.50	0.90	1.20	4.60
9.	Kerala	1.0	0.80	0.30	2.10
10.	Madhya Pradesh	6.0	1.20	3.00	10.20
11.	Maharashtra	4.10	1.20	2.0	7.30
12.	Manipur	0.14	0.10	—	0.29
13.	Orissa	3.60	0.80	1.50	5.90
14.	Punjab	3.00	0.05	3.30	6.35
15.	Rajasthan	2.75	0.40	2.0	5.15
16.	Tamil Nadu	1.50	1.00	1.5	4.00
17.	Uttar Pradesh	12.50	1.20	12.0	25.70
18.	West Bengal	2.31	1.30	2.50	6.11
19.	Union territories & small states, Sikkim, Nagaland, Meghalaya, etc.	0.31	0.40	0.20	0.91
	All India Total	58.50	15.20	39.40	113.10

Source: Sprinkler and Drip Irrigation—Short Term Course Conducted by WAPCOS, New Delhi, 1985.

Development—State-wise

The irrigated area is not evenly distributed over different States. In some states like the Punjab and Tamil Nadu the percentage of area is higher whereas in Madhya Pradesh, Maharashtra, Karnataka, and Gujarat, it is much less. Table 4 gives the total area and the percentage area of irrigation in various States. Rainfall being inadequate, erratic and uncertain over large areas in many States, the development of irrigation has always been one

Table 4. Cropped and irrigated areas in various States (in Mha)

Sl. No. & Name of State	Gross area sown			Gross area irrigated			% of irrigation		
	1970	2000	2025	1970	2000	2025	1970	2000	2025
	2	3	4	5	6	7	8	9	10
1. Andhra Pradesh	13.3	16.7	17.6	4.2	7.3	10.2	32	44	58
2. Assam/Mizoram	2.8	3.5	3.8	0.6	1.5	2.5	21	43	66
3. Bihar	11.1	14.0	16.0	2.7	9.0	13.1	24	64	82
4. Gujarat	9.8	13.1	14.3	1.3	3.9	5.0	13	30	34
5. Haryana	4.9	5.3	5.1	2.2	3.2	3.3	45	62	64
6. Himachal Pradesh	0.9	1.0	1.2	0.2	0.2	0.2	17	20	17
7. Jammu & Kashmir	0.9	1.0	1.0	0.3	0.6	0.7	33	60	67
8. Karnataka	10.9	14.0	14.4	1.4	4.3	5.9	13	31	41
9. Kerala	2.9	3.2	3.2	0.6	1.6	2.6	21	50	81
10. Madhya Pradesh	20.6	26.3	28.4	1.5	5.9	9.1	7	22	32
11. Maharashtra	18.8	32.8	24.6	1.7	5.3	6.5	9	22	26
12. Manipur	0.1	0.3	0.4	0.1	0.2	0.2	40	50	60
13. Meghalaya	0.2	0.3	0.4	—	0.1	0.1	20	25	25
14. Nagaland	0.1	0.2	0.3	—	0.1	0.1	12	35	45
15. Orissa	6.2	10.2	10.5	1.6	4.3	6.7	24	42	64
16. Punjab	5.7	6.0	5.9	4.2	5.0	5.0	75	83	85
17. Rajasthan	16.7	15.4	15.4	2.5	4.4	4.8	15	29	31
18. Tamil Nadu	7.4	9.2	9.5	3.4	4.0	4.0	46	43	42

19. Tripura	0.3	0.3	0.3	—	0.1	0.1	7	25	40
20. Uttar Pradesh	23.2	27.0	28.5	8.4	18.1	24.0	36	67	84
21. West Bengal	7.2	8.1	9.0	1.5	4.4	5.5	21	54	61
22. Union territories	0.5	0.5	0.7	0.1	0.2	0.3	24	40	43
All India	165.1	200	210	38.5	84.0	110	23	42	52

Source: Report of the National Commission on Agriculture, 1976 Part V Resources Development, New Delhi.

of their main primary needs. In spite of the rapid progress made in the last 35 years of increasing the irrigation potential, the entire water resources of the country have as yet not been harnessed. It is expected that by the turn of the century, about 113 Mha of land will be brought under irrigation by using all the harnessable water resources in the country.

Scope

In India, the scope for expanding the area under cultivation is limited. The additional requirements of food and fibre to meet the demand of the increased population have, therefore, come to be met mostly by increasing the yield in unit areas. But this is possible only by providing irrigation since the variability of rainfall is high over the entire country.

The gross irrigated area has increased from 22.6 Mha in 1951 to 71.32 Mha in 1985. But some of the irrigation systems designed to suit the needs then are not able to satisfy present day requirements. The demand for water is increasing day by day for all purposes. The future need of the country can be successfully met by proper planning, development and efficient use of water resources, improving the existing systems to meet the needs of the improved and intensive agriculture, planning conjunctive use of surface and ground water, adopting proper watershed management practices to arrest soil erosion, and developing command areas to ensure rapid and efficient utilisation of irrigation potential. Further, the present practice of irrigation methods have to be oriented towards bringing more area under irrigation from the available water.

Irrigation practices

Water is to be given to the crops whenever there is a need. The method of application depends upon many factors such as topography of the land, characteristics of the soil, type of crop grown, quality and quantity of water, the nature and availability of inputs like labour and energy, economic status of the farmers, etc.

From time immemorial, surface irrigation methods have been followed. The different types of surface irrigation are uncontrolled flooding, controlled flooding, border irrigation, contour border, check basin, furrow, corrugation and basin methods. The quantity of water used by these methods is based on the availability and not in accordance with the requirements of the crops. Though the science of irrigation has developed greatly in the last 20 years in India with reference to soil-water-plant relationship, scheduling of irrigation, water requirements of crops and improved technologies are not widely adopted in the country.

Sprinkler and drip irrigation systems are very popular in many countries in the world like the U.S.A., Australia, Israel, Europe etc. but these have not yet become popular with the farmers in general in India. Many planters in the country, however, for tea, coffee and vegetables are using the sprinkler irrigation system, while coconut and orchard farmers have been using the drip irrigation system for the past five to ten years.

Though large areas are irrigated by advanced methods of irrigation elsewhere, it is not so in India, but the time is ripe to introduce sprinkler irrigation for closely spaced crops like groundnut, cotton, sugar cane, millets and pulses, forage crops and drip irrigation for wide spaced high value crops like coconut, grapes, banana, lime, etc. In order to increase crop production in keeping with the population growth, more area of land, needs to be brought under irrigation. This is possible only by introducing the sprinkler and drip irrigation systems in the coming years, replacing surface methods for certain crops and certain locations.

The sprinkler method is becoming increasingly popular in India in regions of water scarcity where water is insufficient to irrigate the command area by the surface method. In these areas by adopting suitable cropping patterns more area can be brought under irrigation and farm income can be increased substantially. There is no other alternative except introducing this method especially for close spaced crops not only in well irrigated areas but also in tank and canal commissioned areas. Energy constraint should not stand in the way of popularising sprinkler irrigation in the coming years.

CHAPTER 2

Development of Sprinkler Irrigation

The sprinkler or overhead irrigation system consists of conveying water to the field by aluminium or Polyvinyl chloride (PVC) pipes and distributing it over the field under pressure through a system of nozzles. For spraying water under pressure, a booster pump or high speed low discharge pump is necessary. Since the water is conveyed through pipes, the seepage and evaporation losses which are about 20 per cent in well irrigated areas to 50 per cent in canal irrigated areas are eliminated. Further, the system can be designed to distribute the required depth of water uniformly which is not possible in surface irrigation. Since the water application rate is less than the infiltration rate of the soil, there are no runoff losses in this method of irrigation. Even when the soil is too porous and difficult to distribute water uniformly in the surface method, it can be irrigated efficiently by a sprinkler. This method can be used under most climatic conditions where irrigation is feasible.

Though it is more than 75 years since the development of sprinkler irrigation technology in the world took place, little of this method was known or talked about till five or ten years ago in the country. People generally associate sprinkler irrigation with droughts or water scarcity areas. Due to the demand of more water and to bring more area under irrigation, the Government as well as farmers are interested and eager to introduce sprinkler irrigation on a large scale. The development of sprinkler irrigation is still very little and has a long way to go in economising the use of water.

World-wide development

Large scale development of sprinkler irrigation did not get under way till about 1946. However, since then there has been a tremendous development of it, particularly in Europe and in the U.S.A. Though detailed data are not available, it appears that in 1967, about 2.6 Mha were irrigated with sprinkler in Europe. At the end of 1967, there were about 3.0 Mha irrigated with sprinklers in the U.S.A. representing about 17 per cent of the total irrigated area. Sprinkler irrigation accounts for virtually all of the net increase in irrigated area in the U.S.A. between 1960 and the early eighties. The data furnished in 1985 indicated that the area of sprinkler irrigation is about 9.00 Mha out of 25 Mha cultivated, which works out to about 40 per cent of the irrigated area.

Sprinkler irrigation in other countries of the world is under varying degrees of development. In Israel 0.15 Mha representing about 95 per cent of the total irrigated area is by sprinklers. A number of installations have been made with more planning in Tunisia, Libya and Turkey. Experimental units were installed in Taiwan in 1952 for irrigating sugar cane with encouraging results. In Greece and Italy, sprinkler irrigation is widely adopted. Australia has introduced sprinkler irrigation on a large scale in their orchards and for fodder crops in about 50,000 ha. It was estimated in the year 1970, that sprinkler irrigation was adopted in an area of about 6 Mha in the world, which is significant not only from the standpoint of the gain in the world food production, but also because it was accomplished almost entirely by individual farmers with their own financial resources.

Developments in India

Sprinkler irrigation was not a familiar sight to the farming community in India till 1980, though this system is working satisfactorily in the western countries. It is because an average Indian farmer is poor and he cannot afford this system. Further, the problem of getting water except in a few States is not difficult since the rainfall and water potential is abundant. Only during the Sixth Plan period, the irrigation potential was increased to 3 Mha per year and hence it is expected that all water resources will be harnessed by 2000 by bringing about 113 Mha under irrigation which will be about 50 per cent of the total cultivated area at that time. The scarcity of water was realised in States like Tamil Nadu, Karnataka, Haryana, Gujarat and Maharastra, and hence large scale adoption of this method is in progress. Further, the Government of India to conserve water, has taken an initiative to give a subsidy to the farmers to an extent of 20 per cent to 50 per cent in order to popularise this water-saving device. In the early years, plantation farmers used sprinklers for irrigating their coffee, tea, cardamom and other crops raised on sloping hills as supplemental irrigation during non-rainy periods which enhanced the crop yields. The erratic behaviour and failures of monsoon in some places have led the farmers to irrigate their farms and the sprinkler can be adopted for the purpose in undulated and sloping hills and on mountain tracts.

Though detailed statistics are not available, it is estimated that the total gross area under sprinkler irrigation is currently about 0.23 Mha in India out of the total irrigated area of about 71.32 Mha. Indigenous manufacture of the systems started about 20 years ago using foreign designs and technology. Subsequently with research and development, the equipment and production techniques have been modified and adapted to suit Indian conditions, especially in the areas of strength and durability, operating economy and reduced cost for the users.

In India, development work on sprinkler systems has been mainly to reduce the cost of equipment, but not at the expense of durability and to save energy by the introduction of smaller, low pressure sprinklers. Till recently, the most popular piping material has been aluminium but light weight galvanised steel pipe is now being introduced in India. This is much stronger than aluminium and available at a lower cost to the farmer.

The spread and popularisation of the sprinkler with farmers has received significant support from the various schemes involving subsidy of the Central and State Governments. The following are some of them:

a) Integrated Rural Development Programme: In the centrally financed (SFDA, DPAP)—various rates of subsidy are available for small/marginal and medium farmers both individually and on a community/cooperative basis for minor irrigation schemes including sprinkler irrigation. Karnataka Government has taken up an extensive programme providing small farmers with bore-well and sprinklers on a community basis.

b) Centrally sponsored schemes for encouraging irrigation through sprinklers, solar pumps, wind mill pumps, etc. The Government of India contributes 50 per cent of the subsidy with the State Government contributing 50 per cent. Subsidy rates are 50 per cent for small farmers and 20 per cent for others.

c) Centrally sponsored intensive project for increasing production of groundnut. This offers substantial subsidy on the sprinkler system purchased for the irrigation of groundnut.

d) Haryana is operating a highly successful subsidy-cum-loan scheme specifically for sprinkler irrigation since the seventies. A 25 per cent subsidy is provided by the Government and loans are given to the farmers through land development bank and the same is refinanced by NABARD. This type of scheme has now been taken up in Rajasthan, Gujarat, Maharashtra, and Madhya Pradesh.

It is stated that about 10,148 sprinkler sets were installed in Haryana State up to 31.3.1983 at a total estimated cost of Rs. 20 crores. This has resulted in providing irrigation to an additional area of 42,000 ha without any extra water. Further, it was revealed that due to introduction of the sprinkler irrigation system, the gross and net irrigated areas increased to an extent of 83.24 and 66.65 per cent respectively.

Research and experimental findings

The following are some of the useful findings available for introducing sprinkler irrigation:

The Harayana Irrigation Department has reported that saving of water by a sprinkler was seen compared to surface irrigation averaging to 56 per cent in the case of *bajra*, *jowar*, wheat, barley and gram and 29 per cent in the case of cotton. The Punjab Agricultural University has reported a water saving of 42.7 per cent for wheat and 47.5 per cent for maize. The University of Agricultural Sciences, Bangalore, has found that the net irrigated area and cropping intensity were higher when sprinkler irrigation was introduced in the university farm.

A comparative study on sprinkler irrigation in respect of

cotton (MCU. 5) was undertaken with the basin method of irrigation as control at Coimbatore. The cotton *kapas* yield was the highest under sprinkler irrigation (23.3 Q per ha) consuming only 316 mm of water whereas the basin system recorded the lowest yield of 18.5 Q per ha consuming the largest quantity of water (610 mm).

Evaluation of the sprinkler irrigation method on groundnut crop at Bhavanisagar revealed that irrigating the crop to a depth of 3 cm at 0.6 IW/CPE ratio recorded significantly higher pod yield (1878 kg per ha) whereas flow irrigation with 5 cm depth of water at 0.6 IW/CPE ratio recorded 1,557 kg of pods per ha.

Studies on sprinkler irrigation comparing with surface irrigation were made at Madurai on groundnut revealed that sprinkler irrigation was preferable for both yield betterment and water saving. The reason for better yield under sprinkler irrigation with less quantity of irrigation water was that it provided adequate moisture for plant growth, while keeping the soil structure loose and friable, which was conducive to good aeration, root growth and pod formation. Illustrations of sprinkler irrigation, for various crops are shown in the Plates.

Place of Sprinkler Irrigation

The population of India is fast approaching the 750 M mark and is anticipated to become 1000 M by 2000. The estimated food requirements in 2000 would be about 235 MT compared to the 1984-85 production of 153 MT. This wide gap needs to be narrowed by adopting scientific approaches for the efficient management of land and water resources of India.

Though both land for production and water are in short supply, the demand of the former can be obtained by increasing the intensity of cultivation. The average intensity of cultivation in India is only 120 per cent, and only 20 per cent of the cultivated area has two crops in a year. It is possible to increase this to 200-300 per cent provided water is available for cultivation. Therefore, water will be the main constraint for the growth of cultivation in the coming years. To some extent, this constraint could be compensated by introducing sprinkler irrigation. The land and water use in India in 1980-81 is given below:

1) LAND

Geographical area	328 Mha
Total cultivated area	178 Mha
Current total (gross) cropped area	173.20 Mha
Net cropped area	143 Mha
Area sown more than once	30.2 Mha
Gross irrigated area	59.60 Mha
Area irrigated more than once	11.60 Mha

2) WATER
   Total water resource potential   400 MhM
   Utilisable water   95 MhM/100 MhM
   Current utilisation   65 MhM
   Current share of irrigation   61.1 MhM
   Anticipated share of irrigation in
     2000 A.D.   77 MhM
   Ultimate irrigation potential   113 MhM

3) IRRIGATION POTENTIAL
   Total irrigation potential   113 MhM
   Surface   73 Mha
   Ground water   40 Mha

In spite of the large investments made in the irrigation sector and the phenomenal growth of irrigation potential since the year 1951, the returns for the investment in terms of yield as well as finance are disappointing. Irrigated lands should at least yield 3 to 5 t/ha of grains when appropriate water management and other cultural practices are maintained at optimum levels. But the average productivity of irrigated land is about 1.0 to 1.5 t/ha while for the lands irrigated by canal water, it is less than 1.0 t/ha. The major drawbacks are high seepage in channels and fields, insufficient control structures and absence of field channels to irrigate and collect excess water. In order to eliminate or overcome these deficiencies and to increase the area of irrigation, sprinkler irrigation can be introduced even in canal irrigated areas for all crops except rice. In advanced countries like the U.S.A., large areas under a canal command are irrigated by sprinklers. Even in the Nile delta of Egypt canal water is lifted by pumping, to irrigate land which helps to control and monitor irrigation water.

A satisfactory irrigation system means that it gives the correct amount of water to the soil needed to maintain an adequate and constant supply of soil moisture to the root zone of the crop. It should be at a reasonable cost, with minimum waste of water, land, power and labour. Sprinkler irrigation can contribute significantly in this respect. The advantage of sprinklers over surface methods will vary from place to place and time to time. Of course in some situations, surface methods may be more desirable since no two farms have the same requirements. It is necessary to consider all facts for a particular farm to decide which type of irrigation will be most satisfactory.

The benefits of sprinkler method taking into account its use of land and water are given here. This will fall under four categories: (a) Water conservation, (b) Soil conservation, (c) Crop benefits, and (d) Labour benefits.

a) WATER CONSERVATION

It is generally believed that considerable savings will result in going for sprinkler irrigation especially under conditions extremely adverse for the surface method. It the water is conveyed through an unlined earthern channel, in the case of the surface method prevailing in India about 15 to 20 per cent is lost by seepage in garden land condition and 30 to 50 per cent in the case of canal and tank irrigation systems. Further, it is not possible to apply water uniformly in all places in surface irrigation, whereas the system can be designed to give uniform distribution in sprinkler irrigation. Therefore it is possible to increase the area by one and a half to two times by changing from the surface to sprinkler system with the same quantity of water (Fig. 1).

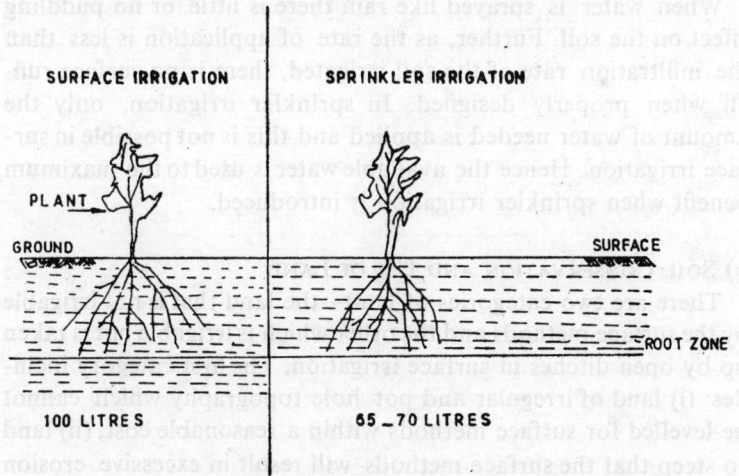

Fig. 1. Comparison of surface and sprinkler irrigation.

F.L. Overly who carried out long range experiments in orchard irrigation with sprinklers in the U.S.A; found that the sprinkler

saved as much as 25 per cent of water. Experiments conducted in various places in India have revealed that the saving of
water varies from 25 to 50 per cent for different crops (Table 5).

Table 5. Experimental results with sprinkler irrigation for
various crops

Sl. No.	Crop	Water saved %	Yield increase v/v
1	Bajra	10-20	Nil
2	Groundnut	25-35	Nil
3	Maize	17	5-10
4	Cotton	20-30	Nil
5	Potato	60	Nil

Source: Sivanappan, R.K. Proceedings of Sprinkler and Drip Irrigation
Seminar, 1984, Delhi.

It is reported that there will be some losses when water is sprayed. Christiansen has estimated that direct evaporation from the
spray itself when normal pressure is applied in a normal wind
velocity condition did not exceed 2 per cent.

When water is sprayed like rain there is little or no puddling
effect on the soil. Further, as the rate of application is less than
the infiltration rate of the soil irrigated, there is no surface runoff when properly designed. In sprinkler irrigation, only the
amount of water needed is applied and this is not possible in surface irrigation. Hence the available water is used to the maximum
benefit when sprinkler irrigation is introduced.

b) SOIL CONSERVATION AND USE OF LAND

There are two categories of land—the land that is not irrigable
by the surface methods and the other which is irrigable but is taken
up by open ditches in surface irrigation. The first category includes: (i) land of irregular and pot hole topography which cannot
be levelled for surface methods within a reasonable cost; (ii) land
so steep that the surface methods will result in excessive erosion
or high cost of application of water; (iii) isolated high areas
which cannot be reached by gravity ditches; and (iv) soil so thin or
porous that excessive losses by deep percolation occur when the
surface methods are used.

In all the above soils, sprinkler irrigation can be used profitably. By introducing this method, there will be no soil erosion problem, no compaction of soil during irrigation, no land levelling required, no land being lost to formation of ditches (as in the case of surface irrigation), and it will control leaching of alkali and other salts. On the whole, it appears that the effect of properly applied sprinkler irrigation is not harmful to the soil and in some ways it is more beneficial than surface irrigation.

c) CROP BENEFITS

Soil moisture is maintained at optimum level by sprinkler irrigation and so higher yields are obtained in some crops. The quality of the products is also good. It helps in providing frost protection. Since the water is sprayed over the crops, it permits cooling of crops. In this system, fertilisers and pesticides can be mixed with water and applied, and hence the efficiency of these inputs for crop production is more when compared to the surface irrigation method.

d) LABOUR BENEFITS

Sprinkler irrigation is automated and since once in six or eight hours the lateral pipes need to be shifted, it reduces labour requirements. But where the land and the farm distribution systems are unusually well adopted to the surface methods, labour costs may be much less than with the sprinkler system.

In India, the emphasis has been on providing the maximum possible area with irrigation facilities. But the optimum utilisation of available water has not received the same emphasis. Not only do losses vary from 30-50 per cent in major irrigation projects, but also large areas are lost due to salinity, alkalinity and water-logging problems in the project. This is not only a national problem, but also a problem of individual farmers. Though much progress has been made in the use of high yielding varieties and fertilisers and pest control measures, the limiting factor is irrigation. Adoption of sprinkler irrigation can give a large boost to production.

Farmers, adopting sprinkler irrigation, assert that their lands are having less infestation of pests and diseases cutting down thereby, the plant protection costs. Furtiliser can also be applied through the sprinkler system which saves on labour. The cumulative effect of these advantages is to ensure that a sprinkler system pays for itself in a few seasons when developing new land, and it may well be less expensive than the surface method, since land levelling is extra, and construction of channels is not needed.

From the above it is clear that sprinkler irrigation has a lot to offer the progressive farmers in increasing production, reducing costs and on a National scale in intensifying the Green Revolution.

DISADVANTAGES

However, there are certain limitations which stand in the way of introducing sprinkler irrigation system on a large scale. One disadvantage is wind. A carefully planned distribution pattern can be completely distorted by it. Wind conditions should be given consideration in the original design of the system to minimise this disadvantage. Ripening soft fruit must be protected from the spray. A stable water supply is needed for the most economical use of the equipment. The water must be clean and free of sand debris and large amounts of dissolved salts.

The sprinkler method usually requires the highest initial investment as compared to surface methods, except where extensive land levelling is necessary for surface irrigation. Power requirements are usually high since sprinklers operate with a water pressure of 0.5 to more than 10 kg/cm². Fine textured soils that have a slow infiltration rate cannot be irrigated efficiently in hot windy areas.

More water is lost by evaporation during sprinkling than with surface flood method. The loss by evaporation will depend on climatic and operating conditions, but it may amount to about 2 to 5 per cent of the water used. On some soils, movement of portable pipes after irrigation may pose a problem. This is true on dry type soils that drain slowly. Orchard irrigation by sprinkling has unique problems. When sprinklers are used under a tree, low hanging branches may interfere with the uniform distribution of water. Where sprinklers are located above the trees, losses due to evaporation and wind interference reduce effectiveness of uniform sprinkling.

CHAPTER 4

Hydraulics

Hydraulics deals with the applications of hydro-mechanics to engineering problems.

In sprinkler irrigation systems, water is pumped from a source (well, river or ponds) through pipes to the sprinklers and then sprayed as uniformly as possible over the crops. To use sprinkler equipment, it is not necessary to understand the complexities or the design, but a working knowledge of how water is pumped and flows through pipes and how it is distributed by sprinklers will help the irrigator and farmer to make full use of his equipment.

Pressure and pipe friction

Pressure of liquid at any point increases directly with the depth of that point and for a given depth the liquid exerts exactly the same pressure in all directions. The vertical depth at any part in a liquid at rest from the free surface is known as pressure head. The unit of pressure is kg/cm², at a height of equivalent liquid column in metres. The advantage of this pressure representation becomes obvious if the term wh (pressure) is divided by the specific weight w then $\dfrac{wh}{w} = h$. The devices designed for measurement of the intensity of hydraulic pressure are based on either by the two fundamental principles of measurement of pressure: firstly, by balancing the column of liquid by the same or another column of liquid; and secondly, by balancing the column of liquid by a spring or dead weight. Water can be conveyed from one portion to another in closed conduits or in open channels over a combination of both. In sprinkler irrigation, it is conveyed through closed conduits under high pressure. The size of the pipe to be used in connection with sprinkler irrigation should be decided in such a way that the most economical pipe is employed which involves minimum outlay of capital with maximum efficiency. The major loss of the pipe in the sprinkler irrigation system is due to friction. The losses depend to a great extent on the type of water whether it is streamline or turbulent. If the Reynolds number is less than 2000, it is called 'laminar' and if more than 2000, it is called 'turbulent'. The Reynolds number is calculated by the formula:

$$R_N = \frac{Vd}{Kr}$$

where R_N = Reynolds No.
V = Velocity of fluid in M/S
d = Diameter of pipe in mm
ν = Kinemetic viscosity of water in M²/S = $\dfrac{\mu}{\rho}$
K = Constant equal to 1000

If expressed as a function of the discharge q

$$Re = \frac{4q}{k\nu\pi D}$$

where q = the pipe discharge expressed in l/h
k = the constant equal to 3600

Three flow regimes may be defined as a function of Re
Laminar flow regime when Re $\leqslant$ 1000
Unstable flow, regime when 2000 < Re $\leqslant$ 4000
Turbulent flow regime when Re > 4000

Fluid friction or friction losses

The frictional losses are heavy in long pipes. Therefore it exerts a tremendous influence on the design and choice of pipes both for main and laterals in the sprinkler system. The frictional losses are calculated by many formulas for various pipes. These are given below:

Darcy-Weisbach equation:

$$hf = fl\, v^2/2\, Dg$$

where hf = head loss when flow occurs in a pipe
l = the distance along which the head loss occurs

v = mean velocity flow
D = diameter of the pipe
f = friction coefficient
g = the gravitational acceleration

The friction coefficient 'f' depends upon 'Re' and the relative roughness of the pipe.

Hazen Williams formula

Although the Darcy-Weisbach equation gives a rational solution of pipe flow problems, various empirical formulas for flow of water in pipes have been developed from laboratory or field data. The best known and most extensively used of these is the Hazen Williams formula.

$$V = 1.32 \, CRh^{0.63} \, S_f^{\,0.54}$$

where V = the velocity
C = a discharge coefficient
R_h = the hydraulic radius
S_f = slope of the energy line

Value of C for various kinds of pipes are given in Table 6.

Table 6. Value of C in Hazen Williams formula

Description of pipe	C
Polyvinyl chloride pipe	150
Extremely smooth and straight	140
Very smooth	130
Smooth wood and wood stave	120
Vitrified	110
Old revetted steel	100
Old cast iron	95
Old pipe in bad condition	60-80
Small pipes badly tuberculated	40-50

DISCHARGE

The quantity of water passing through a given section of pipe, sprinkler nozzle, etc. during a given period of time expressed in cu m per hour or litre per sec.

PRESSURE MEASUREMENT

Pressure in a pipe system can be measured using a Bourdon gauge. Irrigators measure pressure in the field using these gauges as they are simple to use. If this gauge is replaced by a long vertical tube, the water pressure in the pipe would cause water to rise up the tube. The height of this water column is a measure of the pressure in the pipe.

APPLICATION RATE

The rate at which sprinklers apply water when a group of them are operating close together is called the application rate. This is measured in mm/hour. The application rate depends on the size of the sprinkler nozzle, the operating pressure and the spacing between the sprinklers. Increasing the nozzle size or pressure and bringing the sprinklers closer together will increase the application rate. The rate of application should always be less than the rate at which the soil can absorb water (Infiltration rate of the soil).

DROP SIZES

A sprinkler normally produces a wide range of drops with sizes varying from 0.5 mm to 4.0 mm in diameter. Larger drops can damage delicate crops, could break down soil structure, and reduce the infiltration rate. In such cases, sprinklers ejecting small drops should be used to reduce the damage. The range of drop size can be controlled by the size of the nozzle and its operating pressure.

FLOW IN PIPES

Pipes are used to supply water to the sprinklers. Their size, wall thickness and strength depend on the discharge they must carry and the pressure required in the system. It is difficult to design a system that will provide the right pressure for every sprinkler. Pressures vary throughout a pipe system as losses occur due to friction. As the system works under pressure, pipes must be able to withstand high pressures without bursting.

FACTORS AFFECTING PIPE FLOW

As water flows along a pipe, there is a gradual loss of pressure due to friction. Although the inside of a pipe may seem to be very smooth to touch, they can be quite rough hydraulically. This roughness slows down the water flow in the same way as friction slows down an object pushed over a rough surface.

Pressure losses depend not only on the roughness but also on the discharge, the pipe diameter and the pipe length. If the discharge in the pipe increases, the flow velocity also increases and this causes the friction to rise rapidly resulting in much greater pressure loss.

The length of the pipe has a direct effect on the pressure loss. The farther the water has to travel, the more friction will it generate and the greater will be the pressure loss. The relationship between these factors is coupled.

EFFECT OF GROUND ELEVATION ON PRESSURE

Sprinkler systems are often used in areas where the land topography is undulating or sloping steeply. This change in ground elevation will cause changes in pressure in a pipe. This pressure drop will obviously affect the sprinkler performance further up the slope, particularly as pressure is also being lost along the pipe through friction. To avoid this, sprinkler laterals should be laid out level along the ground contour. If this is not possible then an allowance for elevation change must be made in determining the pressure required in the system.

Laying sprinkler laterals on a gently downward slope can be of benefit as the pressure increases. This increase can be used to offset losses that occur from pipe friction. Too much of a slope can cause problems as the pressure rise may become too high resulting in sprinklers operating above their recommended pressure.

PUMPING PRESSURE REQUIREMENTS

The pressure to be supplied to the sprinkler system by pumping must take account of:
   a) recommended pressure at sprinkler;
   b) pressure losses in the mainline/laterals; and
   c) changes in ground elevation.

CHAPTER 5

Sprinkler Equipments

A sprinkler system usually consists of the following parts (Fig. 2):

a) A pumping unit.
b) Tubings—mains, sub-mains and laterals (pipe lines).
c) Couplers.
d) Sprinklers.
e) Other accessories such as valves, bends, plugs and risers.
f) Debris removal equipment.

The cost of the sprinkler system is roughly spread over as follows:

	In percentage
Pumping unit	... 30
Tubings	... 50
Couplers	... 10
Sprinklers	... 5
Other fittings	... 5
Total	100

The description of the various parts are:

a) PUMPING UNIT

Sprinkler irrigation systems distribute water by spraying it over the fields. The water is pumped under pressure to the fields. The pressure forces the water through sprinklers or through perforations or nozzles in pipe lines and then forms spray. Sometimes the slope of the land is sufficient to provide gravity pressure in a pipe line or a central pumping plant is used for a number of sprinkler systems. A high speed centrifugal or turbine

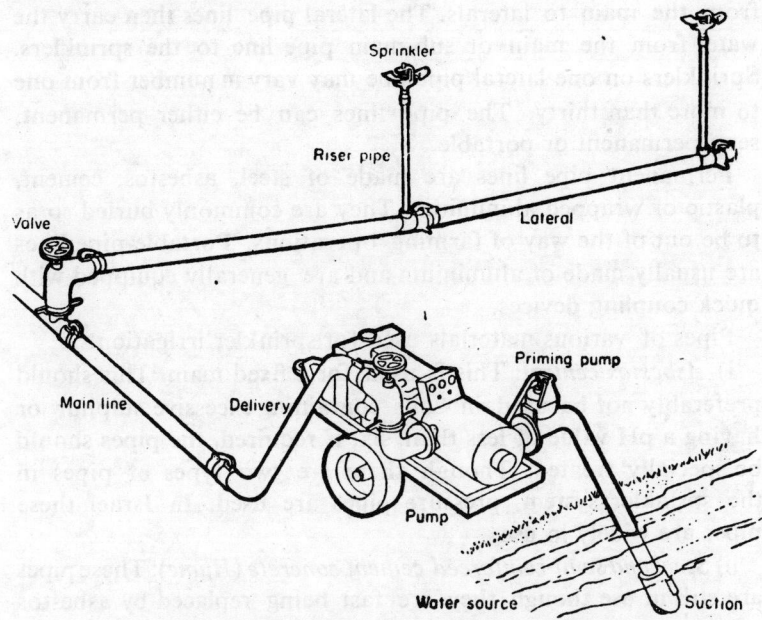

Fig. 2. Components of a sprinkler system.

pump can be installed for operating the system for individual farm holdings. However, the pressure must be provided by a pumping plant for each system. The pumping plant usually consists of a centrifugal or a turbine type pump, a driving unit, a suction line and a foot valve.

A centrifugal pump is generally used when the distance from the pump inlet to the water surface is less than eight metres. Normally centrifugal pumps are used to lift water from irrigation ditches, drainage canals, lakes, ponds, river channels or shallow wells. If the distance to the water surface is more than eight metres or if the water level is fluctuating widely, the use of a turbine pump is recommended. The driving unit may be either an electric motor or an internal cumbustion engine.

b) TUBINGS OR PIPE LINES

Pipe lines are generally of two types—main and lateral. Main pipe lines carry water from the pumping plant to many parts of the field. In some cases sub-main lines are provided to take water from the main to laterals. The lateral pipe lines then carry the water from the main or sub-main pipe line to the sprinklers. Sprinklers on one lateral pipe line may vary in number from one to more than thirty. The pipe lines can be either permanent, semi-permanent or portable.

Permanent pipe lines are made of steel, asbestos, cement, plastic or wrapped aluminium. They are commonly buried so as to be out of the way of farming operations. Portable pipe lines are usually made of aluminium and are generally equipped with quick coupling devices.

Pipes of various materials used for sprinkler irrigation.

i) *Asbestos cement*: This is used for a fixed main. This should preferably not be used in soils containing excessive sulphur or having a pH value of less than six; if required, the pipes should be specially treated. Though there are two types of pipes in the sprinkler system, pressure pipes are used. In Israel these pipes are widely in use.

ii) *Steel and semi-reinforced cement concrete* (*Hume*): These pipes are still in use though they are fast being replaced by asbestos cement. A small diameter lateral hardly carries four to six sprinklers. The only advantage steel pipes have over aluminium is on price, but welded aluminium tubings are available at reasonably comparable prices.

iii) *Aluminium tubing*: This is by far the most preferred tubing in sprinkler irrigation. It is easily portable, and is available in sizes of 5 cm to 25 cm diameter in 3, 6, 9 and 12 m lengths. A tube can be specified according to the manufacturing process as required. It can be drawn, extruded or scum welded. Extrusion entails heavy initial investment and proves economical only if production is on a large scale to offset material and labour costs.

iv) *Polyethylene*: It is now used for this system. For larger diameter (20 to 30 cm) collapsable polyethylene tubings are used as gated pipes. For a sprinkler lateral 3 to 5 cm tubing can be used. The material should lend itself suitably for coiling during shifting and for use as a straight pipe during sprinkling. Suitable material for thin wall tubing for operating pressure ratings of minimum 30 m head, is available. In certain conditions, a combination of aluminium sub-main and polyethylene laterals may prove economical.

c) COUPLERS

Steel couplers are in use with steel pipes. Cast steel couplers are heavy and expensive. A coupler provides connection between two tubings and between tubings and fittings. Essentially a coupler should:

i) provide a reuse and flexible connection;
ii) not leak at the joint under pressure;
iii) automatically drain at no pressure;
iv) be simple and easy to couple and uncouple; and
v) be light, non-corrosive, durable.

It is made of aluminium steel and cast steel and classified as ball and socket, ball only and spherical. The locking device in the coupler may be coil spring, hook, latch, or clamp. A coupler can be plain or with a riser outlet provided for sprinkler connection. The common problems associated with a bolted coupler are leakage of water and the coupler slipping off the tube.

d) SPRINKLERS

Sprinklers may rotate or remain fixed (Fig. 3). Those that rotate can be adapted for a wide range of application rates and spacings. They are effective with pressure of about 10 to 70 m head at the sprinkler. Pressures ranging from 16 to 40 m head are considered the most practical for most farms.

Fixed head sprinklers are commonly used to irrigate small lawns and gardens. Perforated lateral pipe lines are sometimes

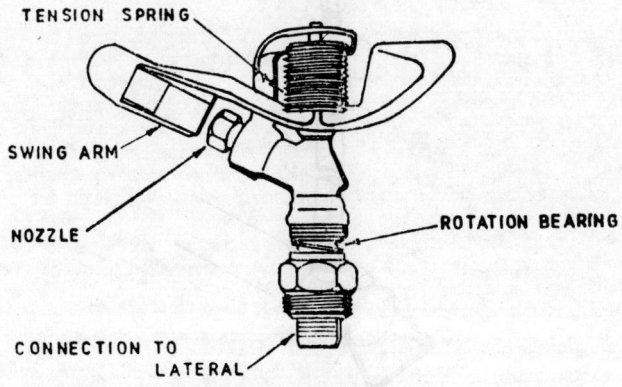

Fig. 3. A good sprinkler.

used as sprinklers. They require less pressure than rotating sprinklers. They release more water per unit of time than rotating sprinklers. Their use should be restricted to soils that have high intake rates. Various factors affect sprinkler performance, such as:

i) *Wind*: Wind is a factor on which we have no control. Hours of high wind velocities (16 km per hour) should be avoided while using a sprinkler. Sprinkler spacing is adjusted to suit the wind conditions. It is generally observed that the wind velocity is less at night and hence the gradual shift to night irrigation.

ii) *Operating pressure*: Sprinklers are available for a very wide range of pressure ratings. The trend is towards medium pressure sprinkling. High pressure sprinklers provide extensive coverage.

iii) *Nozzle*: A sprinkler may be equipped with one, two or three nozzles. A range of nozzle sizes is available for use on a sprinkler. A designer should preferably select a medium nozzle combination so that a wide range of rain is possible from the system by changing nozzles. Plastic nozzles are being developed to replace brass.

iv) *Riser*: A sprinkler should stand above the crop so that its jet is not interrupted by the foliage (Fig. 4). High risers should be

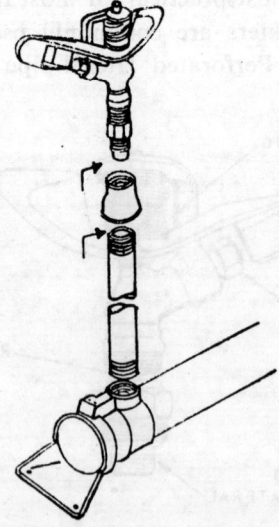

Fig. 4. Risers connected the sprinklers to the lateral.

avoided unless necessary. High risers are used on crops like sugar cane, maize, orchards and they should be adjustable for height. They should not he made of aluminium. Some of the problems met with high risers are:

1) Changing riser height with crop growth, which requires the riser to be in sections.

2) Coupling the riser with another.

3) Riser support to suit varying riser heights.

The riser must be light, capable of easy coupling and uncoupling at all points and easily portable.

e) OTHER ACCESSORIES/FITTINGS

The following are some of the important fittings and accessories used in a sprinkler system.

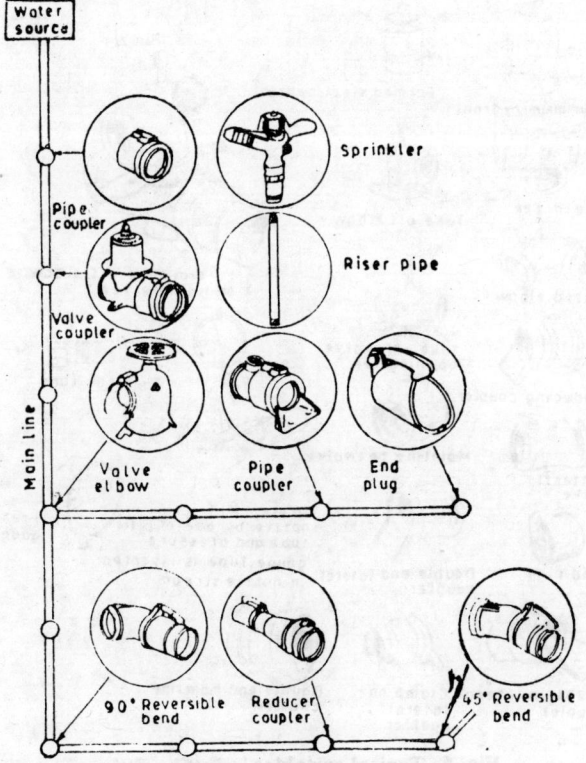

Fig. 5. Pipe specials and accessories.

i) *Water meters*: It is used to measure the volume of water delivered. In India, though irrigation water is normally not measured and users pay according to the area irrigated; this is necessary to operate the system to give the required quantity of water.

ii) Flange, couplings and nipple for proper connection to the pump and suction delivery.

iii) *Pressure gauge*: It is necessary to know whether the sprinkler is working with the desired pressure in order to deliver the water uniformly. A portable gauge-pack with a pitot tube enables an operator to read the sprinkler pressure at the sprinkler nozzle which is in use.

iv) Bends, tees, reducers, elbows, hydrants, butterfly valves, end plugs and crosses are fabricated from aluminium tubings of

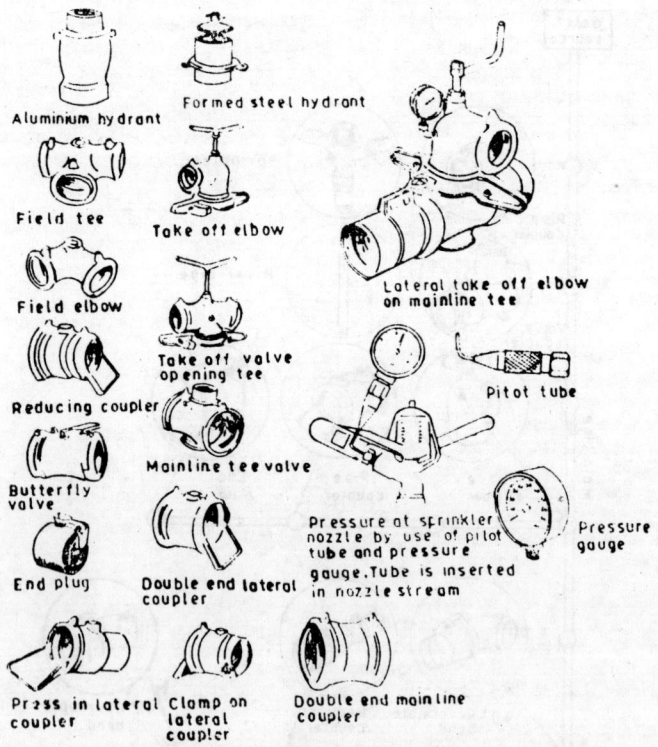

Fig. 6. Typical sprinkler irrigation fittings.

required sizes as shown in Fig. 5. While drawing a bill of material a designer should provide suitable couplers with all fittings.

v) *Fertilisers applicators*: These are available in various sizes. They inject fertilisers in liquid form to the sprinkler system at a desired rate. Continuous flow of water through the applicator is provided by slight difference in pressures at the two tapping points on the sprinkler system—one at the inlet end and the other at the outlet end.

f) DEBRIS REMOVAL EQUIPMENT

The debris removal equipment is needed for most sprinkler systems that obtain water from streams, ponds, canals or other surface supplies. When the water is pumped from wells, this equipment is generally not needed. It is important to keep the sprinkler system clear of sand, weed seeds, leaves, sticks, moss and other trash that may plug the sprinklers.

Typical sprinkler irrigation fittings are given in Figs. 5 and 6.

CHAPTER 6

Types of Sprinkler Systems

Sprinkler systems are classified in the two following major types on the basis of the arrangement for spraying irrigation water:
1) Rotating head or revolving sprinkler system.
2) Perforated pipe system.

1) Rotating head or revolving sprinkler system
This can again be divided into three categories namely:
a) Conventional system/small rotary sprinklers.
b) Boom type and self propelled sprinkler system.
c) Mobile raingun/large rotary sprinklers.

a) ROTARY SPRINKLERS/REVOLVING SPRINKLER/ROTATING
  HEAD/CONVENTIONAL SYSTEM
Small size nozzles are placed on riser pipes fixed at uniform intervals along the length of a lateral pipe (Fig. 7). The lateral pipes are usually laid on the surface of the ground. They may also be mounted on posts above the crop height and rotated through 90° to irrigate a rectangular strip. In rotating type of sprinklers the most common device to rotate the sprinkler head is with a small hammer actuated by the thrust of the water striking against a vane connected to it. This system using many small rotary sprinklers operating together were the first to make sprinkler irrigation popular in the thirties and they are still the most commonly used systems today. The sprinklers operate at low to medium pressures of two to four bar and can irrigate an area of 9-24 m wide and up to 300 m long at one setting. Appplication rates vary from 5 to 35 mm per hour.

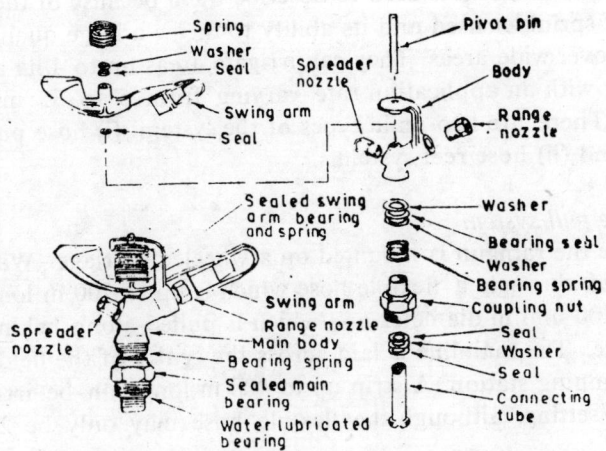

Fig. 7. Components of a rotary sprinkler.

b) BOOM TYPE/SELF-PROPELLED SPRINKLER SYSTEMS

This system employs one boom sprinkler on each lateral. The boom is a nozzled, slowly rotating pipe line which is suspended from a portable tower. Booms sprinklers are moved by towing the towers to the next position along the laterals with a tractor or winch. The large sprinkler irrigates a width of 75 to 100 m depending on nozzle sizes and pressure and is particularly useful for tall crops such as corn and sugar cane, where space at regular intervals is available for manoeuvring the portable towers.

A self-propelled sprinkler consists of a radial pipe line supported at a height of 1.8 to 2.4 m, at intervals of about 30 m on towers mounted on two wheels or a small truck. The radial line is rotated slowly around the first point in the centre of the field by either water pressure actuators or by electric motors at each tower. Conventional sprinklers mounted on the pipe then distribute water to the field as the pipe line is moving. This system covers about 10 to 100 ha and the total capacity range is from 1500 to 4500 lit/min. This type of sprinkler is often used for crops where it is difficult to move sprinkler laterals in the conventional manner.

c) MOBILE RAINGUN/LARGE ROTARY SPRINKLERS

This system operates at high pressure to irrigate large areas.

The term 'raingun' is used to describe them because of the large size of sprinkler used and its ability to throw a large quantity of water over wide areas. They can irrigate areas up to 4 ha at one setting with an application rate varying from 5 to 35 mm per hour. There are two main types of the system: (i) hose pull system; and (ii) hose reel system.

i) *Hose pull system*

Here the raingun is mounted on a wheeled carriage. Water is supplied through a flexible hose which is up to 200 m long and 50 to 100 mm in diameter and which is pulled along behind the carriage. The mainline is laid across the centre of the field from the pumping station. A strip up to 400 m long can be irrigated at one setting, although the flexible hose may only be 200 m long.

The raingun carriage is moved either by a water motor powered from water supply using a piston, or a turbine drive, or an internal combustion engine. The application rate is controlled by the pressure at the raingun. The forward speed of the machine controls the depth of water, which speed varies from 10 to 50 mm an hour. The faster the machine travels, the smaller is the depth of water applied.

ii) *Hose reel system*

The hose reel machine has a raingun mounted on a wheeled carriage. Water is supplied through a more rigid hose than that used for the hose pull although it is still flexible enough to be wound on to a large reel. The hose is used to pull the raingun towards the hose reel positioned at the edge of the field. Machines are available with hose lengths ranging from 200 to 400 metres.

In a layout, the mainline is placed across the centre of the field from the pumping station. The hose reel is placed close to the mainline at the start of the first run and connected to the water supply. The raingun is slowly pulled out across the field by a tractor and the hose allowed to uncoil from the reel. The pump is started and the valve couple slowly opened to start the irrigation. The raingun is slowly pulled back across the field by winding the hose on to the hose reel. Power to drive the hose reel can be provided by a water motor, an internal combustion engine or the power takeoff point on a tractor.

All these revolving sprinkler methods have their advantages and limitations. It is believed that in developing countries where farm labour costs are low and investment capital limited, the 'hand move' system will predominate for many years.

2) Perforated pipe system

This method consists of drilled holes or nozzles along their length through which water is sprayed under pressure. The system is usually designed for relatively low operating pressure (1 kg/sq. cm). The application rate range from 1.25 to 5 cm per hour is for various pressures and spacings. It is therefore limited to use in soils having fairly high intake rates. A layout of Perfo-spray irrigation is given in Fig. 8. There are three types of spraying systems: (a) Stationary, (b) Oscillating, and (c) Rotating.

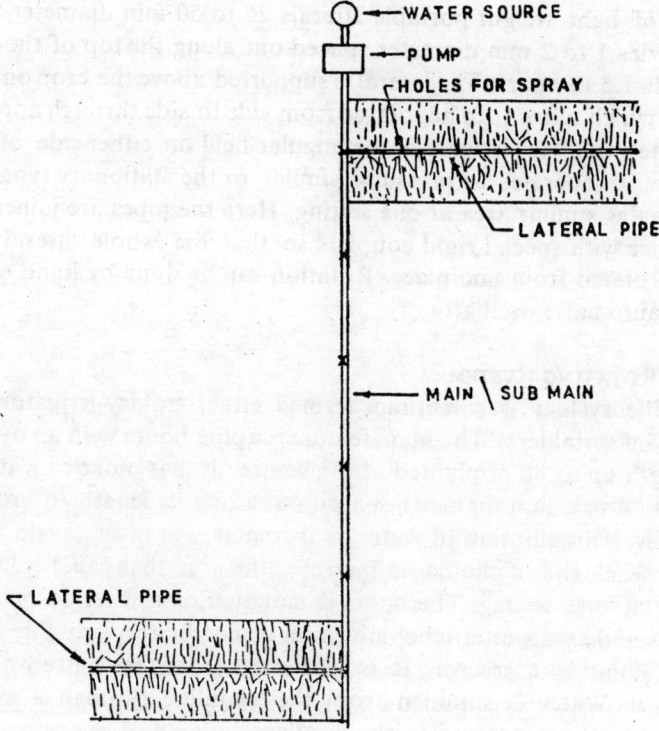

Fig. 8. Perfo-spray irrigation layout.

a) STATIONARY SYSTEM

They consist of light weight portable laterals 50 to 100 mm diameter connected with quick couplers. Small holes of 1 to 2 mm diameter are drilled into the top side of the pipe, so that water sprays in all directions wetting a rectangular area. This operates at pressures between 1.5 and 5 kg/cm² and can irrigate an area of land 5 to 15 m wide and up to 200 m long at one setting. The rate of application varies from 10 to 30 mm/hour.

When irrigating, the mainlines and laterals are laid out and operated in a manner similar to the portable system using rotary sprinklers. Laterals are moved to the field in the same way. This is used for horticultural crops and for nurseries and often for delicate seedlings and plants.

b) OSCILLATING SYSTEM

This is also used mainly for small scale horticulture. It consists of light weight portable laterals 25 to 50 mm diameter with nozzles 1 to 2 mm diameter spaced out along the top of the pipe 0.6 to 1.5 m apart. The lateral is supported above the crop on stands which allow it to be rotated from side to side through approximately 100° to irrigate a rectangular field on either side of the pipe. It operates at pressures similar to the stationary type and irrigates similar area at one setting. Here the pipes are joined together with special rigid couplers so that the whole lateral can be rotated from one place. Rotation can be done by hand or by an automatic oscillator.

c) ROTATING SYSTEM

The system is sometimes termed either 'rotary irrigators' or 'boom sprinklers'. The main feature is a pipe boom with an overall length up to 80 m pivoted at the centre. It has nozzles varying from 4 to 8 mm diameter spaced out along its length to provide an even distribution of water as it rotates. A ranger nozzle fitted into each end of the boom increases the area that can be irrigated at one setting. The boom is mounted on a wheeled carriage powered by an internal combustion engine or through the take-off point in a tractor. It is supported from the centre by steel cables. Water is supplied from the mainline through a swivel joint to the boom and to the nozzles.

Based on the portability, sprinkler systems are classified into

the following types:
1) Portable system.
2) Solid set or permanent system.
3) Semipermanent system.

1) Portable system

The simplest portable system designed is to be moved by hand. It consists of a pump, mainline lateral and rotary sprinklers spaced 9 to 24 m apart. The lateral is usually between 50 mm to 100 mm in diameter so that it can be moved easily. It remains in position until irrigation is complete. The pump is then stopped and the lateral disconnected from the mainline and allowed to drain. It is then dismantled and moved by hand labour to the next point on the mainline and reassembled. As the lateral is connected to the end of the mainline it is also necessary to disconnect sections of the mainline. Usually the lateral is moved between one and four times each day depending on the set time. It is gradually moved around the field until the entire field is irrigated.

In some cases to irrigate large areas, two to three laterals are used. They are connected to the mainline using valve couplers. This allows irrigation to continue while one of the laterals is being moved. In the system described, only the laterals are moved during irrigation while the mainline remains permanently in the same place. In some cases, the entire system including pump and the mainline is moved from field to field. Systems where part or all of the equipments are hand moved regularly, are called portable systems.

Portable systems are the most popular and are used to irrigate a wide range of field and orchard crops, for which capital investment is comparatively low and they are simple to use. As the equipment is to be moved often, labour is required. The equipment is therefore well suited to Indian conditions (Figs. 9, 10 and 11).

At places where labour is inaccessible, or expensive, it may not be possible to use hand move systems. Under these circumstances, mechanical move system employing a variety of devices for moving the lateral line by rotating 'booms' to drag, winches or wheel lines or sprinklers on a rotating tower arrangement may be used. In such cases, pipes must be strong, and rigid couplings used to carry the loads.

Another labour saving system is the flexible lateral system

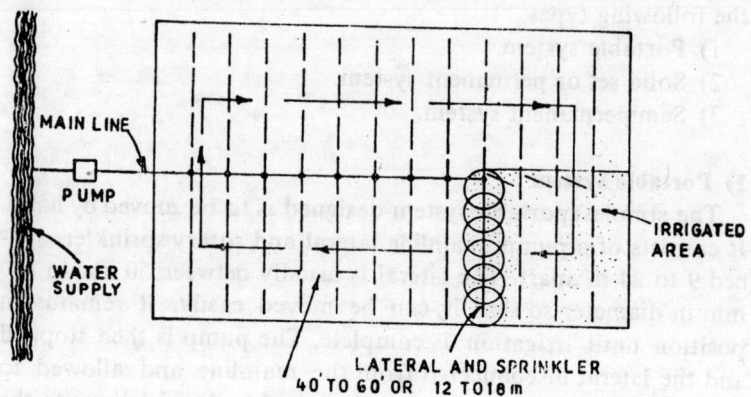

Fig. 9. Portable system using one lateral (hand moved).

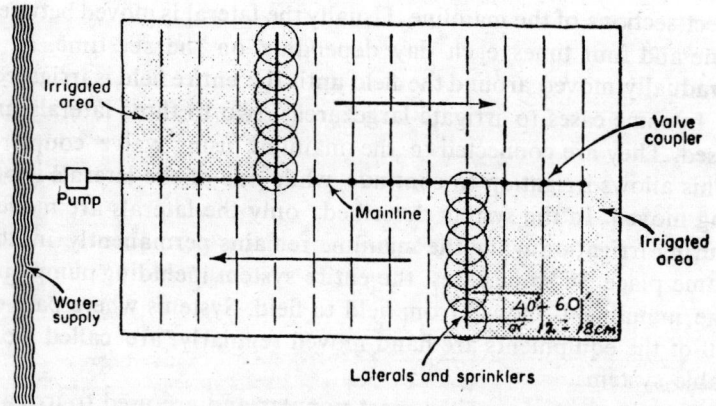

Fig. 10. Portable system using 2 laterals (hand moved).

which can be wound up on to a drum at the end of each irrigation. The rotary sprinklers are connected to the lateral at intervals on special frames. They lie flat when being coiled and pop up vertically when irrigating.

2) Solid set or permanent system

When sufficient laterals and sprinklers are provided to cover the entire area to be irrigated, there is no need for moving the equipments from place to place. This system is termed "solid set or permanent system" (Fig. 12).

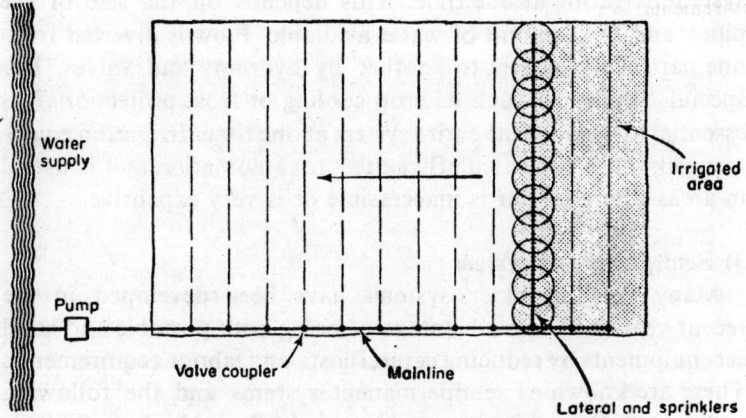

Fig. 11. Alternative layout for a hand moved portable.

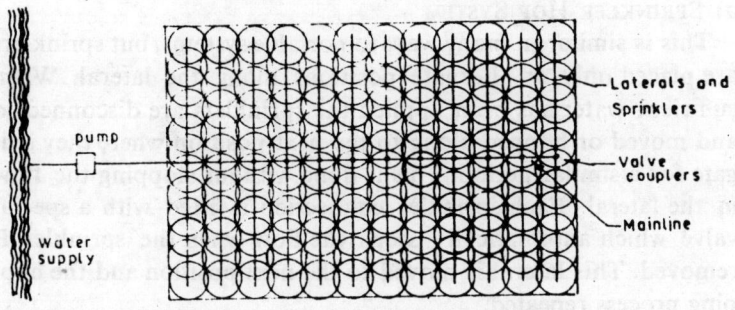

Fig. 12. Solid set or permanent system.

This type is used on permanently irrigated areas and for relatively high value crops such as orchards, vine yards, pastures and nurseries. The purpose of this system is to reduce labour costs and eliminate moving of lateral lines especially where tree foliage is heavy. Fully permanent systems are not many, because of the relatively high installation costs, even though in many cases, these costs would be more than offset by saving in labour costs. The most commonly used materials for buried pipe are asbestos-cement, concrete pressure pipe, and coated steel. In recent years, plastic, fibre glass, and aluminium alloy pipe designed for underground use have been successful.

Most solid and permanent systems have only a part of the

system irrigating at one time. This depends on the size of the pipes and the amount of water available. Flow is diverted from one part of the system to another by hydrants and valves. For special conditions such as crop cooling or frost protection, it is essential to operate the entire system at one time. Irrigation equipment like this is particularly suited to automation and is useful in areas where labour is inaccessible or is very expensive.

3) Semipermanent system

Many new sprinkler systems have been developed in the recent years with the advantage of being both portable and solid set equipments by reducing capital costs and labour requirements. These are known as semipermanent systems and the following systems are the most commonly used: (a) Sprinkler hop, (b) Pipe grid, and (c) Hose pull.

a) SPRINKLER HOP SYSTEM

This is similar in many ways to portable systems, but sprinklers are placed only at alternate positions along the lateral. When sufficient water has been applied the sprinklers are disconnected and moved or hopped along to the next position where they irrigate for a similar period. This is done without stopping the flow in the lateral. Each sprinkler connection is fitted with a special valve which automatically stops the flow when the sprinkler is removed. This lateral is moved to the next position and the hopping process repeated.

b) PIPE GRID SYSTEM

This is similar to the solid set or permanent system. Small diameter laterals about 25 mm are used to keep system costs low. The pipes are laid out over the entire field and they remain in place throughout the irrigation season, thus eliminating movements of pipes. Two sprinklers are connected to each lateral, one near the top, the other half way down. When sufficient water has been applied each sprinkler is disconnected and moved along the lateral to the next position. This is repeated until the whole field has been irrigated.

c) HOSE PULL SYSTEM

This is used for under tree irrigation for orchard crops and other row crops. The mainline and laterals are usually permanently installed either on or below the ground surface. Small diameter plastic hoses supply water from the lateral to one or two rotary sprinklers. The hose length is about 50 m because of friction losses in the pipe. This reduces a number of laterals and provides great flexibility in irrigation. A sprinkler can be easily moved where it is needed and can be adjusted to compensate for the distortion of spray patterns caused by wind. It is cheap, but problems can arise with plastic hoses.

CHAPTER 7

Basic Data Needed/Sprinkler Design Factors

To design a sprinkler irrigation system, some basic data are needed. What is the information required and in what way it is useful to design the system are discussed in this chapter.

1) WATER QUALITY

The quality of water is an important factor to reckon with before introducing this system. If the water is not good for irrigation and has unsatisfactory chemical qualities, the project has to be dropped. The water should be tested by an agronomist or a soil scientist for suitability to grow crops, and also to ensure that it will have no corrosive effect on the pipe and equipment.

2) WATER SUPPLY

The next factor to be considered is whether sufficient water is available (either from a well or from any other source) to meet the peak irrigation requirements in the planned area. The peak annual or seasonal water use for any crop is a function embracing many factors including climate, growing season, soil, topography, wind movement and irrigation practices.

3) MAPS AND OTHER DATA

a) A scaled map of the area to be irrigated should be prepared showing roads, building and other obstructions, water source and possible pump locations.

b) Contour lines can be shown in the map. If slopes are uniform the contour interval can be from 5 to 10 m.

c) Soil types should be noted.

d) If different crops are to be grown, boundaries of each should be marked.

e) Power available for the pump or pressure head in case of gravity may be noted.

4) AVAILABLE SPRINKLER EQUIPMENT

The availability of the equipment in the planned regions including service facilities must be known before designing a system. Engineers cannot design a system to a specific installation without complete knowledge of the available equipment.

5) ESTIMATING CONSUMPTIVE USE (CU)

The Blanney-Criddle formula to estimate monthly and seasonal crop water requirements can be used (Tables 7 and 8).

$$U = KF = \text{sum of } k_f \qquad \text{where}$$

$U = CU$ of crop for any period,

$F = $ Sum of monthly CU factors for the period.

$k = $ Empirical CU crop coefficient for a month,

$t = $ Mean monthly temp. in F°.

$P = $ Monthly percentage of daylight hours in the year for the latitude.

$$f = \frac{t \times p}{100}$$

$u = kf = $ monthly CU in inches

In metric units

$u = kp \, (0.457 \, t + 8.13) = $ monthly CU in mm.

$t = $ Mean monthly temp. in C°.

6) OTHER DESIGN FACTORS

In additon to the crop water requirements the following factors should also be taken into account: (i) the amount of effective rainfall; (ii) irrigation efficiency; and (iii) leaching requirements.

i) *Effective rainfall*: It is the amount of rain which falls during the irrigation season and is adequate to meet consumptive use requirements, but does not include excess water which goes into deep percolation or surface runoff.

ii) *Irrigation efficiency*: Since water is delivered through pipes, there are no conveyance losses. Irrigation efficiency is therefore

Table 7. Seasonal consumptive-use crop coefficients (K)
for irrigated crops (the USA)

Crop	Length of normal growing season or period[1]	Consumptive-use coefficient (K)[2]
Alfalfa	Between frosts	0.80 to 0.90
Bananas	Full year	.80 to 1.00
Beans	3 months	.60 to .70
Cocoa	Full year	.70 to .80
Coffee	Full year	.70 to .80
Corn (Maize)	4 months	.75 to .85
Cotton	7 months	.60 to .70
Dates	Full year	.65 to .80
Flax	7 to 8 months	.70 to .80
Grains, small	3 months	.75 to .85
Grain, sorghums	4 to 5 months	.70 to .80
Oilseeds	3 to 5 months	.65 to .75
Orchard crops:		
Avocado	Full year	.50 to .55
Grapefruit	Full year	.55 to .65
Orange and lemon	Full year	.45 to .55
Walnuts	Between frosts	.60 to .70
Deciduous	Between frosts	.60 to .70
Pasture crops:		
Grass	Between frosts	.75 to .85
Ladino white clover	Between frosts	.80 to .85
Potatoes	3 to 5 months	.65 to .75
Rice	3 to 5 months	1.00 to 1.10
Soybeans	140 days	.65 to .70
Sugar beet	6 months	.65 to .75
Sugar cane	Full year	.80 to .90
Tobacco	4 months	.70 to .80
Tomatoes	4 months	.65 to .70
Truck crops, small	2 to 4 months	.60 to .70
Vineyard	5 to 7 months	.50 to .60

[1]Length of season depends largely on variety and time of year when the crop is grown. Annual crops grown during the winter period may take much longer than if grown in the summertime.

[2]The lower values of (K) for use in the Blanney-Criddle formula, U = KF, are for the more humid areas, and the higher values are for the more arid climates.

Source: Sprinkler Irrigation Guidebook, USAID, Washington, D.C., 1969.

Table 8. Monthly percentage of daytime hours of the year

FOR LATITUDES 0° TO 65° NORTH OF THE EQUATOR

Latitude North	Jan.	Feb.	Mar.	Apr.	May	June	July	Aug.	Sept.	Oct.	Nov.	Dec.
1	2	3	4	5	6	7	8	9	10	11	12	13
65°	3.45	5.14	7.90	9.92	12.65	14.12	13.66	11.25	8.55	6.60	4.12	2.64
64°	3.75	5.30	7.93	9.87	12.42	13.60	13.31	11.15	8.58	6.70	4.35	3.04
63°	4.01	5.40	7.95	9.83	12.22	13.22	13.02	11.04	8.60	6.79	4.55	3.37
62°	4.25	5.52	7.99	9.75	12.03	12.91	12.79	10.92	8.50	6.86	4.72	3.67
61°	4.46	5.61	8.01	9.71	11.88	12.63	12.55	10.84	8.55	6.94	4.89	3.93
60°	4.67	5.70	8.05	9.66	11.72	12.39	12.33	10.72	8.57	7.00	5.04	4.15
59°	4.81	5.78	8.05	9.60	11.61	12.23	12.21	10.60	8.56	7.07	5.09	4.31
58°	4.99	5.85	8.06	9.55	11.44	12.00	12.00	10.56	8.56	7.13	5.13	4.55
57°	5.14	5.93	8.07	9.51	11.32	11.77	11.87	10.47	8.54	7.19	5.27	4.69
56°	5.29	6.00	8.10	9.45	11.20	11.67	11.69	10.40	8.52	7.25	5.54	4.89
55°	5.39	6.06	8.12	9.41	11.11	11.53	11.59	10.32	8.51	7.30	5.62	5.01
54°	5.53	6.12	8.15	9.36	11.00	11.40	11.43	10.27	8.50	7.33	5.74	5.17
53°	5.64	6.19	8.16	9.32	10.88	11.31	11.34	10.19	8.52	7.38	5.83	5.31
52°	5.75	6.23	8.17	9.28	10.81	11.13	11.22	10.15	8.49	7.40	5.94	5.43
51°	5.87	6.25	8.21	9.26	10.76	11.07	11.13	10.05	8.48	7.41	5.97	5.46
50°	5.98	6.32	8.25	9.25	10.69	10.93	10.99	10.00	8.44	7.43	6.07	5.65
48°	6.13	6.42	8.22	9.15	10.50	10.72	10.83	9.92	8.45	7.56	6.24	5.86

(Contd.)

1	2	3	4	5	6	7	8	9	10	11	12	13
46°	6.30	6.50	8.24	9.09	10.37	10.54	10.66	9.82	8.44	7.61	6.38	6.05
44°	6.45	6.59	8.25	9.04	10.22	10.38	10.50	9.73	8.43	7.67	6.51	6.23
42°	6.60	6.66	8.28	8.97	10.10	10.21	10.37	9.64	8.42	7.73	6.63	6.39
40°	6.73	6.73	8.30	8.92	9.99	10.08	10.34	9.56	8.41	7.78	6.73	6.53
38°	6.87	6.79	8.34	8.90	9.92	9.95	10.10	9.47	8.38	7.80	6.82	6.66
36°	6.99	6.86	8.35	8.85	9.31	9.83	9.99	9.40	8.36	7.85	6.92	6.79
34°	7.10	6.91	8.36	8.80	9.72	9.70	9.88	9.33	8.36	7.90	7.02	6.92
32°	7.20	6.97	8.37	8.72	9.63	9.60	9.77	9.28	8.34	7.93	7.11	7.05
30°	7.30	7.03	8.38	8.72	9.53	9.49	9.67	9.22	8.34	7.99	7.19	7.14
28°	7.40	7.02	8.39	8.68	9.46	9.38	9.58	9.16	8.32	8.02	7.27	7.27
26°	7.49	7.12	8.40	8.64	9.37	9.30	9.49	9.10	8.32	8.06	7.36	7.35
24°	7.58	7.17	8.40	8.60	9.30	9.19	9.41	9.05	8.31	8.10	7.43	7.46
22°	7.76	7.22	8.41	8.57	9.22	9.12	9.31	9.00	8.30	8.13	7.50	7.56
20°	7.73	7.26	8.20	8.52	9.14	9.02	9.25	8.95	8.30	8.19	7.58	7.88
18°	7.88	7.26	8.40	8.46	9.06	8.99	9.20	8.81	8.29	8.24	7.67	7.89
16°	7.94	7.30	8.42	8.45	8.98	8.98	9.07	8.80	8.28	8.24	7.72	7.90
14°	7.08	7.39	8.43	8.44	8.90	8.73	8.99	8.79	8.28	8.28	7.85	8.04
12°	8.08	7.40	8.44	8.43	8.84	8.64	8.90	8.78	8.27	8.28	7.85	8.05
10°	8.11	7.40	8.44	8.43	8.81	8.57	8.84	8.74	8.26	8.29	7.89	8.08
8°	8.13	7.41	8.45	3.39	8.75	8.51	8.77	8.70	8.25	8.31	7.89	8.11
6°	8.19	7.49	8.45	8.39	8.73	8.48	8.75	8.69	8.25	8.41	7.95	8.19
4°	8.20	7.58	8.46	8.33	8.65	8.40	8.67	8.63	8.21	8.43	7.95	8.20
2°	8.43	7.62	8.47	8.22	8.51	8.25	8.52	8.50	8.20	8.45	8.16	8.42
0°	8.49	7.67	8.49	8.22	8.49	8.22	8.49	8.49	8.19	8.49	8.22	8.49

FOR LATITUDES 0° TO 50° SOUTH OF THE EQUATOR

Latitude South	Jan.	Feb.	Mar.	Apr.	May	June	July	Aug.	Sept.	Oct.	Nov.	Dec.
1	2	3	4	5	6	7	8	9	10	11	12	13
0°	8.49	7.67	8.49	8.22	8.49	8.22	8.49	8.49	8.19	8.49	8.22	8.49
2°	8.55	7.71	8.49	8.19	8.44	8.17	8.43	8.44	8.19	8.52	8.27	8.55
4°	8.64	7.76	8.50	8.17	8.39	8.08	8.20	8.41	8.19	8.56	8.33	8.65
6°	8.71	7.81	8.50	8.12	8.30	8.00	8.19	8.37	8.18	8.59	8.38	8.74
8°	8.79	7.84	8.51	8.11	8.24	7.91	8.13	8.32	8.18	8.62	8.47	8.84
10°	8.85	7.86	8.52	8.09	8.18	7.84	8.11	8.28	8.18	8.65	8.52	8.90
12°	8.91	7.91	8.53	8.06	8.15	7.79	8.08	8.26	8.17	8.67	8.58	8.95
14°	8.97	7.97	8.54	8.03	8.07	7.70	7.08	8.19	8.16	8.69	8.65	9.01
16°	9.09	8.02	8.56	7.98	7.96	7.57	7.94	8.14	8.14	8.76	8.72	9.17
18°	9.18	8.06	8.57	7.93	7.99	7.50	7.88	8.90	8.14	8.80	8.80	9.24
20°	9.25	8.09	8.58	7.92	7.83	7.41	7.73	8.05	8.13	8.83	8.85	9.32
22°	9.36	8.12	8.58	7.89	7.74	7.30	7.76	8.03	8.13	8.86	8.90	9.38
24°	9.44	8.17	8.59	7.87	7.60	7.24	7.58	7.99	8.12	8.89	8.96	9.47
26°	9.52	8.28	8.00	7.81	7.56	7.07	7.49	7.87	8.11	8.94	9.10	9.61
28°	9.61	8.31	8.61	7.79	7.49	6.99	7.40	7.85	8.10	8.97	9.19	9.74
30°	9.69	8.33	8.63	7.75	7.43	6.94	7.30	7.80	8.09	9.00	9.24	9.80
32°	9.76	8.36	8.64	7.70	7.39	6.85	7.20	7.73	8.08	9.04	9.31	9.87
34°	9.88	8.41	8.65	7.68	7.30	6.73	7.10	7.69	8.06	9.07	9.38	9.99
36°	10.06	8.53	8.67	7.61	7.10	6.59	6.99	7.59	8.06	9.15	9.51	10.21

(Contd.)

1	2	3	4	5	6	7	8	9	10	11	12	13
38°	10.14	8.61	8.68	7.59	7.03	6.46	6.87	7.51	8.05	9.19	9.60	10.34
40°	10.24	8.65	8.70	7.54	6.96	6.33	6.73	7.46	8.04	9.23	9.69	10.42
42°	10.39	8.72	8.71	7.49	6.85	6.20	6.60	7.39	8.01	9.27	9.79	10.57
44°	10.52	8.81	8.72	7.44	6.73	6.04	6.45	7.30	8.00	9.34	9.91	10.72
46°	10.68	8.88	8.73	7.39	6.61	5.87	6.30	7.21	7.98	9.41	10.03	10.90
48°	10.85	8.98	8.76	7.32	6.45	5.69	6.13	7.12	7.96	9.47	10.17	11.09
50°	11.03	9.06	8.77	7.25	6.31	5.48	5.98	7.03	7.95	9.53	10.32	11.30

Source: Sprinkler Irrigation Guidebook, USAID, Washington, D.C., 1968.

the application efficiency which is expressed as the ratio of the amount applied to the root zone and to the amount discharged by the sprinklers. This varies from 60-75 per cent depending on evaporation losses and management of the system. It is important that a simpler selection and spacing be properly made to achieve the suggested efficiency.

iii) *Leaching requirements*: If the water or soil is saline, the salt has to be removed from the root zone of the crop. The net amount of leaching water required is after allowing for effective off-season rainfall and deep percolation losses during the irrigation season. Based on the application efficiency the gross amount to be applied can be calculated. It should be emphasised along with the leaching requirements, that proper and adequate drainage must be provided to carry off the leaching water and deep percolation losses.

7) SOILS

The soil and its characteristics are essential to know and it should include the following:

a) Intake rates to determine the optimum rate at which soil will absorb water (Table 9). The minimum application rate for most sprinklers is about 5 mm per hour to obtain good distribution efficiency when operated under favourable climatic conditions.

b) Water holding capacity of soil: Various textures have inherent abilities to retain water. When the maximum is reached, it is called field capacity and any additional amount is an economic loss. A general rule is that the moisture level within the root zone should not fall below 50 per cent of the available soil moisture. It is also desirable to bring the moisture level at or near to field capacity. Tables 10 and 11 can be used to determine the field capacity of various types of soils for effective root depth of the various crops.

c) Peak period consumptive use: The peak daily crop use is needed to determine the minimum capacity requirements for the sprinkler system. Different crops may have their peak rates of use at different times. Therefore, it is important to carefully analyse the individual crop requirements. The local data on peak requirements can be used. When data are not available, the values given in Tables 12, 13 and 14 may be used with the required adjustments

Table 9. Suggested maximum application rates for sprinklers
for average soil and slope

Soil texture and profile	0-5% slope cm/hr	5-8% slope cm/hr	8-12% slope cm/hr	12-16% slope cm/hr
1. Coarse sandy soil to 2 m	5.10	3.75	2.54	1.27
2. Coarse sandy soil over more compact soils	3.75	2.54	1.90	1.02
3. Light sandy loams to 2 m	2.54	2.03	1.50	1.02
4. Light sandy loams over more compact soils	1.90	1.27	1.02	0.76
5. Silt loams to 2 m	1.27	1.02	0.76	0.51
6. Silt loams over more compact soils	0.76	0.63	0.38	0.25
7. Heavy textured clays or clay loams	0.38	0.25	0.20	0.15

Source: Sprinkler Irrigation Guidebook, USAID, Washington, D.C., 1968.

Table 10. Range in available moisture holding capacity of soils
of different texture

	mm of water per cm of depth
1. Very coarse texture—very coarse sands	0.33 to .58
2. Coarse texture—coarse sands, fine sands and loamy sands	0.62 to 1.05
3. Moderately coarse texture—sandy loams	1.05 to 1.46
4. Medium texture—very fine sandy loams, loams and silt loams	1.25 to 1.92
5. Moderately fine texture—clay loams, silty clay loams and sandy clay loams	1.46 to 2.10
6. Fine texture—sandy clays, silty clays and clays	1.35 to 2.10
7. Peaks and mucks	1.66 to 2.50

for local weather and crop conditions. The lower values are for semi-arid and the higher values are for arid type climate. The peak requirement is needed only for design purposes and is not necessarily the criteria for operating the system. The user may be guided by soil moisture and crop condition in determining the frequency and amount of water to apply during the crop growing period. There are various instruments for determining soil moisture. However, feel and appearance method is one practical way to estimate the amount of soil moisture (Table 14).

Table 11. Depth to which crops will extract available soil moisture from deep, well-drained soils

	Depth of moisture extraction cm		Depth of moisture extraction cm
Alfalfa	180	Sugar beets	120
Asparagus	180	Sweet potatoes	120
Deciduous orchards	180	Carrots	90
Grapes	180	Eggplant	90
Hops	180	Peas	90
Sorghums, grains	180	Peppers	90
Sudan grass	180	Squash (summer)	90
Tomatoes	180	Sweet corn	90
Corn (field)	150	Table beets	90
Flax	150	Beans (bush)	60
Melons	150	Cabbage	60
Small grains	150	Pasture	60
Artichokes	120	Potatoes	60
Beans, lima	120	Spinach	60
Citrus orchards	120	Strawberries	60
Cotton	120	Lettuce	30
		Onions	30

Source: Sprinkler Irrigation Handbook, USAID, Washington, D.C., 1968.

Table 12. Peak period average daily consumptive use rates (u_p) as related to estimated actual monthly use (u_m)

Net irrigation application		Computed peak monthly consumptive use rate (u_m) in inches[1]*																
Inches	cm	4.0	4.5	5.0	5.5	6.0	6.5	7.0	7.5	8.0	8.5	9.0	9.5	10.0	10.5	11.0	11.5	12.0
		Peak period daily use rate (u_p) in inches per day*																
1.0	2.54	.15	.18	.20	.22	.24	.26	.28	.31	.33	.35	.37	.40	.42	.44	.46	.49	.51
1.5	3.81	.15	.17	.19	.21	.23	.25	.27	.29	.32	.34	.36	.38	.41	.43	.45	.47	.50
2.0	5.08	.15	.16	.18	.20	.23	.25	.27	.29	.31	.33	.35	.37	.39	.41	.44	.46	.48
2.5	6.35	.14	.16	.18	.20	.22	.24	.26	.28	.30	.32	.34	.36	.39	.41	.43	.45	.47
3.0	7.52	.14	.16	.18	.20	.22	.24	.26	.28	.30	.32	.34	.36	.38	.40	.42	.44	.46
3.5	8.79	.14	.16	.18	.19	.21	.23	.25	.27	.29	.31	.33	.35	.37	.39	.41	.44	.46
4.0	10.16	.14	.15	.17	.19	.21	.23	.25	.27	.29	.31	.33	.35	.37	.39	.41	.43	.45
4.5	11.43	.14	.15	.17	.19	.21	.23	.25	.27	.29	.31	.33	.35	.37	.39	.41	.43	.45
5.0	12.70	.13	.15	.17	.19	.21	.23	.25	.26	.28	.30	.32	.34	.36	.38	.40	.42	.44
5.5	13.97	.13	.15	.17	.19	.21	.22	.24	.26	.28	.30	.32	.34	.36	.38	.40	.42	.44
6.0	15.24	.13	.15	.17	.19	.20	.22	.24	.26	.28	.30	.32	.34	.36	.38	.40	.41	.43

[1]Based on the formula $u_p = 0.034\ u_m^{1.09}\ I^{-0.09}$ where

u_p = Average daily peak period consumptive use in inches.

u_m = Average consumptive use for the peak month in inches.

I = Net irrigation application in inches.

*To convert to centimeters multiply by 2.54.

Source: Sprinkler Irrigation Guidebook, USAID, Washington, DC., 1968.

Table 13. Peak use daily requirements
(range in values for Western United States)*

Crop	cm/day	Crop	cm/day
Alfalfa	.68 to .89		
Corn	.35 to .63	Citrus	.33 to .53
Cotton	.68 to .79	Deciduous	
Flax	.63 to .71	fruits	.56
Grain	.38 to .53	Pasture	.74 to .81
Grapes	.48	Potatoes	.63 to .74
Orchards		Sugar beets	.51 to .63
Avocadas	.48	Tomatoes	.51
		Truck crops	.51 to .66

*Note: Above data from the U.S.D.A., Agricultural Research Service and Soil Conservation Service. Sprinkler application rate must be based on field application efficiency.

Table 14. Guide for judging how much moisture is available for crops

Available soil moisture remaining	Feel or appearance of soil and moisture deficiency				
		Loamy sand	Sandy loam	Loam and silt loam	Clay loam or silty clay loam
	1	2	3	4	5
0 to 25 per cent		Dry, loose, single grained, flows through fingers	Dry, loose, flows through fingers	Powdery, dry, sometimes slightly crusted but easily broken down into powdery condition	Hard, baked, cracked, sometimes has loose crumbs on surface
Moisture deficiency (mm/cm)		0.75 to 0.57	1.07 to 0.83	1.66 to 1.24	1.76 to 1.33
25 to 50 per cent		Appears to be dry, will not form a ball with pressure[1]	Appears to be dry, will not form a ball[1]	Somewhat crumbly but holds together from pressure	Somewhat pliable, will ball under pressure[1]
mm/cm		0.58 to 0.38	0.833 to 0.53	1.25 to 0.833	1.33 to 0.91
50 to 75 per cent		Appears to be dry, will not form a ball with pressure	Tends to ball under pressure but seldom holds together	Forms a ball somewhat plastic, will sometimes slick slightly with pressure	Forms a ball, ribbons out between thumb and forefinger
mm/cm		0.35 to 0.16	0.52 to 0.25	0.83 to 0.42	0.90 to 0.45

1	2	3	4	5
75 per cent to field capacity (100 per cent)	Tends to stick together slightly, sometimes forms a very weak ball under pressure	Forms a weak ball, breaks easily, will not slick	Forms a ball is very pliable, slicks readily if relatively high in clay	Easily ribbons out between fingers, has slick feeling
mm/cm	0.16 to 0.00	0.25 to 0.00	0.41 to 0.00	0.45 to 0.00
At field capacity (100 per cent)	Upon squeezing, no free water appears on soil but wet outline of ball is left on hand 0.00	Upon squeezing, no free water appears on soil but wet outline of ball is left on hand 0.00	Upon squeezing, no free water appears on soil but wet outline of ball is left on hand 0.00	Upon squeezing, no free water appears on soil but wet outline of ball is left on hand 0.00

[1]Ball is formed by squeezing a handful of soil very firmly.

Source: Sprinkler Irrigation Guidebook, USAID, Washington, 1968.

CHAPTER 8

Design and Layout of the System

General principles

Before proceeding with the sprinkler layout and system design, some general principles or rules for good design should be considered. The objective of the overall design is to secure a system that will provide a satisfactory uniformity of distribution with a minimum annual operating cost, including depreciation, power and labour costs. In the design procedure, various alternatives might be compared to determine the most economical combination of capital costs, power costs and labour costs. Consideration should be given to long term changes in these costs and in changing cropping patterns.

a) Under most conditions, minimum costs and desirable operating conditions can be assured when the source of the water supply is as near the centre of the area to be irrigated as possible. If the source of supply is to be a well, it should be located at the centre of the farm. If a mainline is to be used to supply the sprinkler laterals, it should be placed through the centre of the area. This will result in minimum friction losses in laterals, small pipe sizes and generally more uniform application of water.

b) On sloping lands, it is good practice to place the mainlines on the slope and to place the laterals along the contour or downhill slope to minimise pressure variations in the pipelines. The pressure variations in the laterals can be up to 20 per cent of the design sprinkler pressure. With uniform sprinkler nozzles, the variations in sprinkler discharge will then be limited to 10 per cent of the design discharge. When water is to be carried upward from the source of supply to the mainline, provision should be made to regulate the pressure at the laterals to the design pressure.

c) When a system is designed for a part of a farm, consideration should be given to the possibility of expanding the system to cover the entire area.

d) The design should be such that the operation of the system will result in minimum interference with other farm operations.

e) For irregular fields the design should minimise the variations in the number of sprinklers that would be operating at any given time.

f) Proper safety devices for the power unit and pumping plant must be provided which will infuse an automatic shut down of the pump in case of overheating of motor or engine or failure of water supply in the design.

g) Lateral lines should be located at right angles to the prevailing wind direction wherever possible.

Mainlines and lateral layout

Mainlines or sub-mains when used, should usually be up and down the predominant land slopes. When laterals are laid down a slope, the mainline will often be located along the ridge laterals sloping downward from each side. Changes in pipes should be made along the mainline for pressure control. A mainline should be located so that laterals can be rotated in a split line operation thereby minimising the labour for hauling lateral pipe back to the starting point in the plains and hills (Fig. 13). Typical sprinkler layouts are given in Figs. 14 and 15.

After completing the layout of mainlines and laterals usually it is necessary to make some adjustments in one or more of the following:

Number of sprinklers operating.

Water application rate.

Sprinkler discharge.

Spacing of sprinklers.

Operating time per day.

Total operating time per irrigation.

Total capacity.

As the designer gains experience, he can foresee these

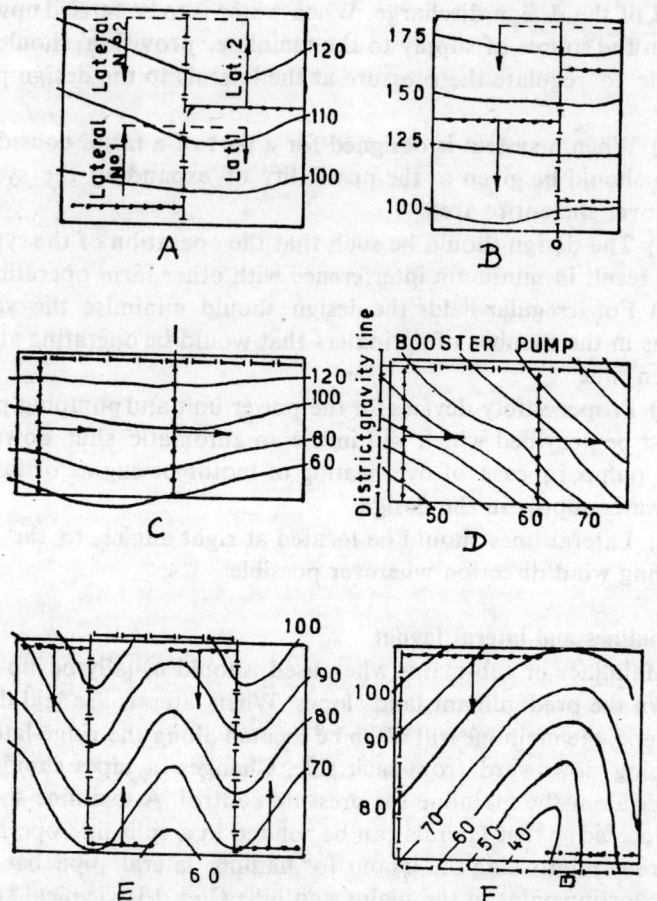

Fig. 13. Layouts of sprinkler systems showing effects of topography. A—Layout on moderate, uniform slopes with water supply at centre. B—Layout illustrating use of odd number of laterals top rovide required number of operating sprinklers. C—Layout with gravity pressure where pressure gain approximates friction loss and allows running lateral downhill. D—Layout illustrating area where laterals have to be laid downslope to avoid wide pressure variation caused by running laterals upslope. E—Layout with two mainlines on ridges to avoid running laterals uphill. F—Layout with two mainlines on the sides of the area to avoid running the laterals uphill.

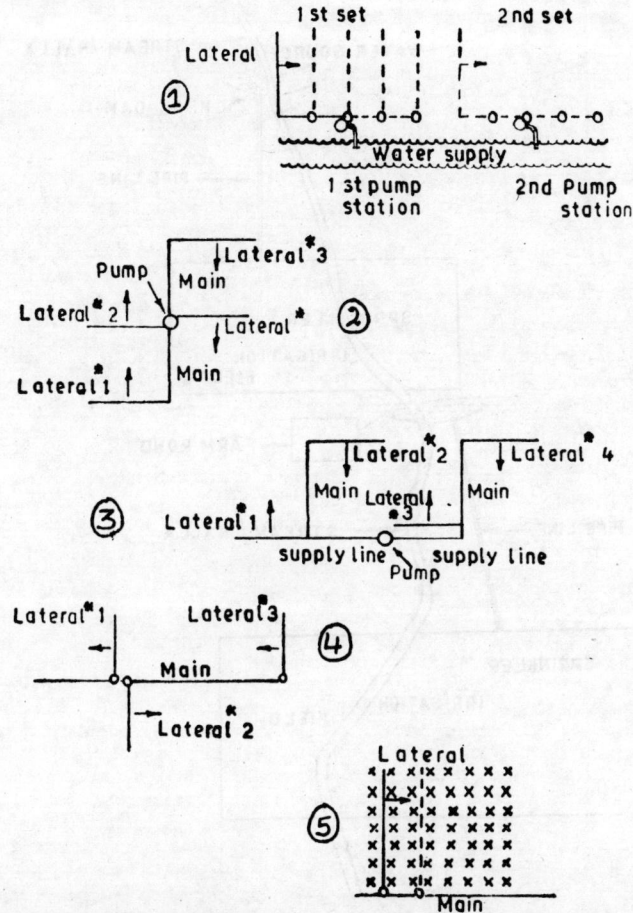

Fig. 14. Typical sprinkler layouts. 1—Fully portable using short mainline to permit 5 locations of lateral from each pump station. 2—Water at centre. Mainline can be on surface or buried. 3—Rectangular shape field. 4—Distribution of odd number of laterals with main through centre line of field. 5—Orchard irrigation with low trajectory sprinklers and alternate tree row spacing. On the following irrigation shift lateral to irrigate in adjacent tree rows.

adjustments before making the layout.

The design of the sprinkler system includes the most feasible layout of the mainlines, the layout and spacing of laterals and the required size of the laterals and mainlines. A definite plan of

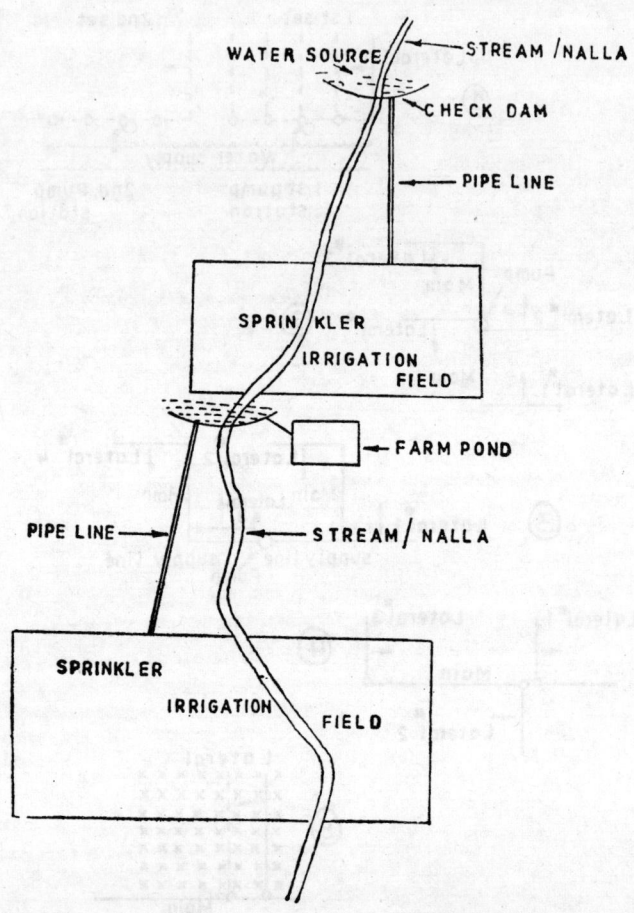

Fig. 15. Typical layout for storing and distribution of water for
sprinkler irrigation in hills and slopy lands.

operation which will insure irrigating the entire area within the
permissible interval between irrigations during the period of
maximum evapotranspiration must be selected. Under normal
conditions sprinkler systems are designed to be operated continuously day and night. The initial cost will be minimum when
it is designed to be operated continuously to meet the maximum
demand. Usually the most satisfactory operating schedule involves
either one or two moves of the sprinkler laterals per day. For
light application (50-60 mm) and especially on soils with fairly

high intake rates, two moves per day are usually satisfactory. For greater application and for soils with low intake rates, one move per day is more desirable. For any given layout, the number of laterals required to cover the area in the permissible irrigation interval depends on the spacing of the laterals which can be readily determined. Under most conditions several laterals supplied by central mainline will prove most economical.

Sprinkler design

A sprinkler irrigation system is specially designed in order to achieve high efficiencies in its performance and economy and also to suit the conditions of a particular site. There are specific steps involved in the planning and design of a sprinkler system which are shown diagrammatically in Fig. 16.

AN EXAMPLE TO ILLUSTRATE METHODOLOGY OF A FARM UNIT SPRINKLER DESIGN

It must be recognised that only the principles of design can be illustrated as it is very seldom that one set of conditions will be duplicated on another farm. The author wishes to emphasise this and to again stress the importance of having at hand all the agronomic and climatological data that are possible for him to obtain prior to doing the design exercise. For the purpose of this example a single rectangular area was chosen with known data from a specified location.

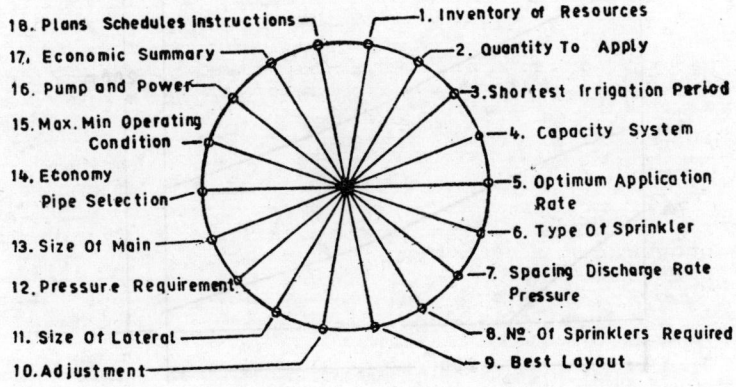

Fig. 16. The 18 steps in design of a sprinkler system.

Step 1: Inventory of resources

Location : Semi-arid type region.

Soil : Moderately coarse. Sandy loam to 2 m, with natural drainage.

Topography : As noted in sketch (Fig. 17).

Water supply : From a well adequate for irrigation.

Water quality : Less than 0.5 gm/litres total dissolved solids (500 parts/million).

Power source : Electricity—3 phase.

Crop : Groundnut (other crops may be grown in rotation).

Climate : Arid and with no effective rainfall during the peak growing period.

Wind : As shown in Fig. 17.

Step 2: Water to apply each irrigation

Referring to Table 10, the total available moisture holding capacity for the soil on this farm ranges from 1.05 to 1.46 mm/cm of depth (1.25″ to 1.75″ per foot). Using an average value of 1.25 mm/cm (1.50″/ft.) and a moisture extraction depth of

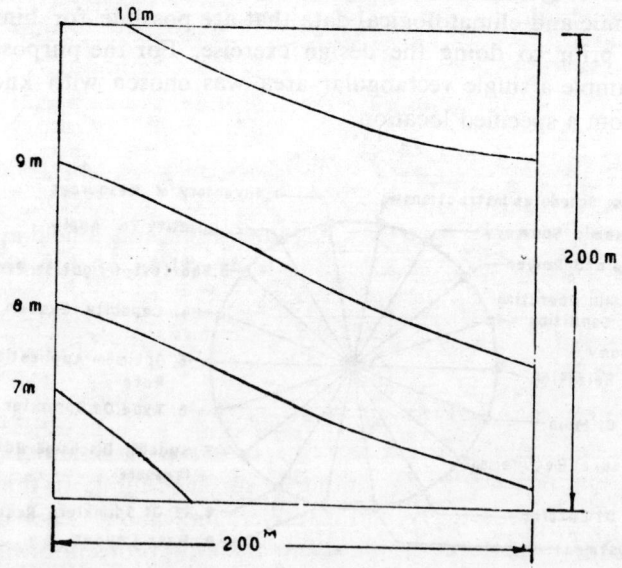

Fig. 17. Dimensions and elevations of field to be irrigated with sprinkler.

60 cm (2') the total available moisture is about 7.5 cm (3"). On the basis of not using more than 50 per cent of this moisture, the maximum amount to be applied at each irrigation is about 3.75 (or) 4 cm.

Step 3: Shortest interval/rotation cycle

This is the time allowable between successive irrigations, commonly called the rotation cycle, during the peak-consumptive use of the crop. Using the value of 0.53 cm/day (for groundnut crop) the rotation cycle becomes $\frac{3.75}{0.53}$ or seven days.

Step 4: Capacity of system

The pump capacity or system requirement is calculated by the equation

Metric	English
$Q = \dfrac{A \times D \times 27.8}{F \times H \times E}$	$Q = \dfrac{A \times D \times 453}{F \times H \times E}$

where

Q — litres/sec of pump	— GPM
A — Area in hectares	— acres
D — Depth application in cm	— inches
F — Irrigation interval in days	— days
H — Hours/day operation	— hours
E — Field application efficiency (fraction)	— fraction
27.8 — Conversion factor	453 — conversion factor

On the basis of 12 hours per day operation and a field application efficiency of 0.80, the capacity required of a pump is

$$\text{Metric } Q = \frac{4 \times 3.75 \times 27.8}{7 \times 12 \times 0.8}$$
$$= 6.2 \text{ l/sec.}$$

Step 5: Optimum water application rate

This is determined by the soil type, crop and slope and is the application rate without puddling or surface runoff. In this example, the slope is fairly uniform averaging about 2 per cent, and as there may be periods where the crop protection will be negligible the maximum application rate is 1.5 cm per hr. Sprinkler selection and spacing must be so as not to exceed this rate.

Step 6: Type of sprinkler to be selected

As previously discussed the sprinkler selection is the key and it will determine the balance of the design work.

Most sprinkler systems are designed for 'round the clock' or continuous operation which will permit the minimum size of pump, main and laterals. Lateral moves generally are on the basis of two or three moves per 24 hours for most crops. This will also be governed by the depth of application required and application rate. However, this is not possible in Indian conditions and hence the design is made accordingly.

Step 7: Sprinkler spacing, nozzle discharge and operating pressure

First make an optimum arrangement on the main and lateral in respect of the topography and operation and then by "cut and try" determine the spacing and sprinkler discharge. In this example, we know the field must be irrigated in a minimum of seven days. If the farm operator prefers two moves, then not more than 14 moves are possible in the case of one lateral or seven moves with two laterals.

Mainlines should always, if possible, run up and down the predominant slope in order to provide the best control for lateral pressures. Laterals should be at or near 90° angle to the main and run across the slope. Also, it is desirable to have the lateral at or near 90° to the prevailing wind. These may not be compatible in which case the slope consideration should take precedence.

We also have certain basic limitations on maximum move intervals, which for the typical intermediate pressure sprinkler is 18 m (60'). Now make a trial layout of the main with the above factors in mind. In our example the location of the main would ideally be through the centre of the field and with 18 m intervals for lateral moves, there will be a total of 11 lateral sets by locating the first and last positions 9 m (30') from the edge of the field. If 15 m (50') move intervals were selected there would be 13 sets.

With the management plan to move the lateral twice a day the field could be covered with two laterals equal to half the width of the field in seven days using the 15 m interval. Thus either arrangement is satisfactory in respect of the time element.

Step 8: Number of sprinklers required

The next step is to select sprinkler spacings and test them for application rates. Here again it is necessary to establish a maximum spacing of the sprinklers on the lateral which for intermediate pressure sprinklers is generally accepted under the given wind conditions at 12 m (40'). Closer spacings are acceptable (and preferably in more windy conditions) so we will "test out" 9 m (30') and 12 m (40') sprinkler spacing. In our example, this becomes 11 or 8 sprinklers respectively by locating the first and last sprinkler 5 m from the mainline and the edge of field.

We therefore have four possible move-spacing selections: 18 × 12 m; 18 × 9 m; 15 × 12 m; 15 × 9 m. Application rates in each case are determined by the formula:

$$\text{cm per hour} = \frac{\text{litre/sec of sprinkler} \times 360}{\text{Spacing} \times \text{move in m}}$$

In this example these become

Move spacing	No. of Sprinklers	Sprinkler size l/sec	Sprinkler size GPM	Application rate cm/hr	Application rate in/hr
18 × 12 m	8	0.8	12.5	1.33	0.50
18 × 9 m	11	0.58	9.1	1.29	0.49
15 × 12 m	8	0.8	12.5	1.60	0.60
15 × 9 m	11	0.58	9.1	1.54	0.58

Referring back to step 5, we find that 1.50 cm per hour is the maximum application rate, for this soil type. All the various selections are perhaps sufficiently close, however a good selection be the 15 × 9 m arrangement as it has the advantage of a low application rate plus the fact that in windy conditions, the closer sprinkler spacing on the lateral will permit more overlap and better distribution efficiency as compared to the 18 × 12 m arrangement.

The next step is to refer to the sprinkler manufacturers catalogue to select the nozzle, pressures and the wetted diameters to meet these requirements. With the 0.58 1/sec sprinkler discharge (9.10 gpm) the nozzle combination is 1/2 cm × 1/4 cm (3/16" × 7/64") for a typical double nozzle sprinkler operating at 3.2 kg/cm² (45 pSI) with a wetted diameter of about 30 m (100').

Step 9: Layout of system

We have completed the basic design and it is well to check at this time to determine if any adjustments are necessary.

Water to apply	— 3.75 cm
Peak water requirement	— 0.53 cm/day
Max. irrigation interval	— 7 days
Pump capacity required at assumed 80 per cent field application efficiency	— 6.2 1/sec.

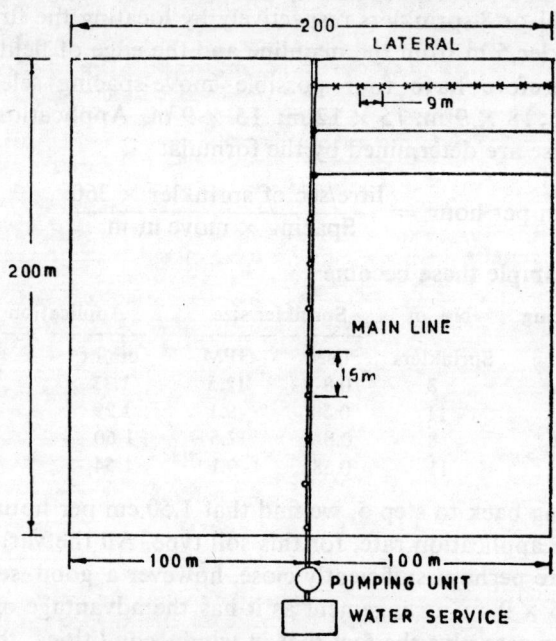

Area **4 ha/10 Acres**
Lateral Spacing **15 m/50′**
Sprinkler Spacing **9 m/30′**

Fig. 18. Layout of sprinkler system for 4 ha (10 acres) land.

Design (see Fig. 18)
Sprinkler spacing — 15 × 9 m (50′ × 30′)
Type of sprinkler — Intermediate pressure
Mainline — 200 m (660′) with lateral outlet each 15 m (50′)
Laterals — 95 m (315′) with 11 sprinklers spaced each 9 m (20′)
Sprinkler capacity — 0.58 1/sec
Application rate — 1.50 cm/hr

Hours per lateral set — six hours
Total gross application — 4.5 cm
Net application (80 per cent eff.) — 3.75 cm
Rotation — seven days with two lateral sets per 12 hours
Daily equivalent moisture replacement over rotation period —
  0.53 cm

Step 10: Adjustments

The above design meets all the requirements. However, it is to be noted that the system must be operated continuously during the peak use period in order to maintain the soil moisture level.

Step 11: Selecting size of lateral

In the interest of economy and ease of handling, the minimum diameter pipe should be selected consistent with good sprinkler performance. A difference of 20 per cent in pressure results in about 10 per cent difference in nozzle discharge and this maximum pressure variation between the first and last sprinkler on a lateral has been generally accepted world wide as an acceptable design criteria.

There have been developed numerous pre-engineered tables logarithmic charts, monographs and special fractions in slide rules to expedite lateral line selection. The design engineer is encouraged to use these for the purpose of methodology. The friction loss and multiple outlet connection factor method is used herein. The step by step procedure follows:

1) Select a given size of pipe.

2) Assume the flow is through the entire length without sprinklers and from Table 15 determine the friction loss.

3) Multiply loss (2) by the correction factor use (Table 16) corresponding to the number of sprinklers on the lateral.

4) To (3) add the elevation if the lateral goes uphill or subtract the drop if the lateral goes downhill.

5) Compare (4) with the allowable 20 per cent loss. If it is approximately the same then your selection is correct; otherwise, select another diameter pipe and repeat the procedure.

In our example, we have selected 3.2 kg/sq. cm (45 psi) as the average design sprinkler pressure which will be at mid-point on the lateral. A 20 per cent variation is 0.64 kg/sq. cm (9 psi) (from first to last sprinkler) or converted to elevation is 6.4 m or 20.8′.

Table 15. Friction head loss in irrigation pipes

Friction loss in metres per 100 m in lateral lines of portable aluminium pipe with couplings (based on Scobey's formula and 9 m pipe lengths) (adopted from Dr A.M. Michael).

Flow Lit/Sec	Diameter of Pipe				
	5 cm Ks 0.34	7.5 cm Ks 0.33	10 cm Ks 0.32	12.5 cm Ks 0.32	15 cm Ks 0.32
1	2	3	4	5	6
1.26	0.32				
1.89	2.53				
2.52	4.49	0.565	0.130		
3.15	6.85	0.858	0.198		
3.79	9.67	1.21	0.280		
4.42	12.90	1.63	0.376	0.122	
5.05	16.7	2.10	0.484	0.157	
5.68	20.8	2.63	0.605	0.196	
6.31	25.4	3.20	0.738	0.240	0.099
7.57		4.84	1.04	0.339	0.140
8.83		6.09	1.40	0.454	0.188
10.10		7.85	1.80	0.590	0.242
11.36		9.82	2.26	0.733	0.302
12.62		12.00	2.76	0.896	0.37
13.88		14.4	3.30	1.07	0.443
15.14		16.9	3.90	1.26	0.522
16.41		19.7	4.54	1.47	0.608
17.67		22.8	5.22	1.70	0.700
18.93		25.9	5.96	1.93	0.798
20.19		29.3	6.74	2.18	0.904
21.45		32.8	7.56	2.45	1.02
22.72		36.6	8.40	2.74	1.13
23.98		40.6	9.36	3.03	1.26
25.24		44.7	10.30	3.34	1.38
26.50			11.30	3.66	1.51
27.76			12.3	4.0	1.66
29.03			13.4	4.35	1.80
30.29			14.6	4.72	1.95
31.55			15.8	5.10	2.12
34.70			18.9	6.12	2.52
37.86			22.2	7.22	2.98
41.00			25.9	8.40	3.46
44.17			29.8	9.68	3.99
47.32			33.8	11.0	4.54

1	2	3	4	5	6
50.48				12.5	5.15
53.63				14.0	5.78
56.79				15.6	6.44
59.94				17.3	7.14
63.10				19.0	7.86

For 6 metres pipe lengths, increase values in the table by 7 per cent and for 12 m lengths decrease by 3 per cent.

Table 16. Factor (F) for computing friction loss in a line with multiple outlets

Outlet number	Value of F	Outlet number	Value of F
1	1.0	18	0.373
2	0.634	19	0.372
3	0.528	20	0.371
4	0.480	21	0.370
5	0.451	22	0.369
6	0.433	23	0.368
7	0.419	24	0.367
8	0.410	25	0.366
9	0.402	26	0.365
10	0.396	27	0.364
11	0.392	28	0.363
12	0.388	29	0.362
13	0.384	30	0.360
14	0.381	35	0.359
15	0.379	40	0.357
16	0.377	50	0.355
17	0.375		

The other conditions are 11 sprinklers each discharging 0.58 l/s (9 gpm) with 95 m of lateral. The lateral locations in this example permit some "downhill" and some level placement.

Let us go through the various steps:

1) Trial size selected 7.5 cm (3").

2) Friction loss through line without outlets using flow of (6.2 1/sec (approx. 100 gpm)

$$3.2 \times 3.15 = 10.08 \text{ feet or } 3.05 \text{ m.}$$

3) Correction factor for 11 sprinklers 0.392.

4) No elevation correction required.

5) Total net loss is 1.2 m (3.95′) compared to 6.4 m (20.7′) allowable.

The use of 7.5 cm (3″) pipe obviously would give excellent operation characteristics. Also if the above tests were applied to 5 cm (2″) pipe, the net loss would be over 8.1 m (27′). The operator would perhaps prefer to use all 7.5 cm (3″). However to illustrate the selection of two pipe diameters in the laterals the following procedure is used:

1) Arbitrarily select some proportion of 7.5 m (3″) and 5 cm (2″). Assume that 32 m (105′) of 7.5 cm (3″) and 63.00 m (210′) of 5 cm (2″) were used.

2) Calculate the loss for the 63 m of 5 cm lateral with seven sprinklers and 4.42 lps (70 gpm) flow (flow of 4.42 lps/70 gpm used as table does not give 40.3 1ps/64 gpm).

$$\text{Loss} = 12.9 \times 2.1 \times 0.419 = 11.3' \text{ or } 3.45 \text{ m.}$$

3) Assume all of flow (6.2 lps) was for the entire length 95 m (315′) using 7.5 cm pipe and 11 sprinklers.

$$\text{Loss} = 3.2 \times 3.15 \times .392 = 4' \text{ or } 1.2 \text{ m}$$

4) Assume the flow 4.42 lps (70 gpm) of the 5 cm section 63 m (210′) with seven sprinklers was in the larger diameter 7.5 cm pipe.

$$\text{Loss} = 1.63 \times 2.1 \times .419 = 1.44' \text{ or } 0.44 \text{ m.}$$

5) Actual total lateral loss with the sum of items (2) + (3) minus item (4) or 3.45 + 1.20 − 0.44 = 4.27 or 4.3 m.

From the above calculation it is seen that a slightly higher proportion of a 5 cm pipe could be used. From a practical operating standpoint, however it is believed that this short lateral, one diameter (7.5 cm) would be more desirable as there would be an appreciable saving in the first cost and the convenience of handling only one size lateral, would soon offset the slightly higher first cost, also would permit expansion to longer laterals if the irrigated area was increased. For systems requiring laterals two sizes are sometimes justified, although most designers prefer to maintain one diameter whenever possible.

Step 12: Pressure required at lateral

Thus far only the midway or 'average design pressure' has been noted. The friction loss along a lateral of uniform size decreases rapidly with three-fourth of the total loss occurring in the first half of the line. To calculate the end sprinkler pressure subtract the pressure to equivalent of one-fourth of the lateral loss from the average design pressure to obtain the pressure at first sprinkler. In our example, the end sprinkler would operate at about 3 kg/sq. cm (44.5 psi) and the beginning sprinkler at about 3.25 kg/sq. cm (46.3 psi). For all practical purposes, there would be equal pressures along the lateral. Pressure increases to compensate for the sprinkler rise pipe will be taken into account later in calculating the system design pressure.

Step 13: Determining required size of mainline pipe

In contrast to lateral design there are no specific standards as to the amount allowable for mainline pipe losses. Hence it is a question of having a "reasonable loss" dictated by either the economics in first cost of the pipe or sometimes in pump-power unit selection. Designers usually consider as "reasonable" a mainline loss of about 3 m (10') for small systems and up to about 12 m (40') for large systems. Also when the system is 'powered' by gravity fall to select sizes to permit proper pressure for the sprinklers.

In our example, the mainline is about 200 m (660'), with a flow of 6.2 1/sec (100 gpm). If a 7.5 cm aluminium pipe were used there would be a total maximum loss of about 7.6 m (25') and with a 10 cm (4″) pipe a loss of 1.8 m (5.9'). It would be logical to use a 10 cm (4″) pipe as it would reduce the pressure requirement by about 1.8 kg/cm² (8 psi) and the saving in power would soon pay for the larger size pipe.

The ultimate objective is to arrive at a selection that results in the lowest annual water-application cost. Various methods and formulas have been evolved to arrive at an "economic-balance"; however, for the purpose of this publication the following guidelines are suggested:

1) Prepare a flow diagram, locating the mainline and the position of the lateral line (or lateral lines) where the "peak" friction loss will occur.

2) Note on this diagram the flow in each section of the mainline.

3) Calculate the friction loss in each section of the mainline

where there are changes in flow. Assumed pipe sizes will be used, starting with the large size at pump and usually changing pipe sizes where there is either a decrease in flow (by sub-main takeoff or lateral) or in case of a long simple mainline at some point to keep the friction loss to a reasonable figure.

4) Calculate the total dynamic head which includes the total losses of the system, the peak pressure requirement and the elevation difference from the water level to the high point in the field.

5) Refer to the pump-power unit catalogues and pump curves and determine the size of pump and power unit required. At this point you may wish to reexamine your pipe selection. If the Total Dynamic Head (T.D.H.) is between two pump-power sizes, perhaps by reducing the mainline friction loss by 10 to 20 per cent you can safely use a smaller size unit or conversely you may find, you have more reserve power than needed and perhaps a change to a slightly higher friction loss can be made.

Step 14: Economy of pipe selected

To make our economic analysis it is necessary to consider not only the pipe sizes but also the particular type of mailing pipe to use (steel, aluminium, plastic, pressure type concrete or cement asbestos) and whether it is to be surfaced or buried, with the outlet valves and other required fittings. The design engineer should have at hand current price lists of available materials. He should also know the future of the planned system, whether the size of the irrigated area is to be increased in the foreseeable future, if the cropping plan may be changed later that will affect the decision of surface or buried lines, etc. Once these data and information are obtained and carefully analysed the mainline and other components can be established and the economics of the chosen system can be determined.

Step 15: Determine operating conditions

In order to determine the size of the pumping unit, list the maximum operating conditions. For our example these are:

	English	Metric
1) Pump capacity	100 GPM	6.2 lps
2) Minimum lateral pressure		
3.2 kg/cm² (46.5 psi at mainline) 107 ft.		33 m

3) Height of sprinkler riser
   (groundnut)   2 ft.   0.6 m
4) Maximum mainline loss   6 ft.   1.8 m
5) Elevation water source to
   high point   46 ft.   14 m
6) Minor losses (elbows, val-
   ues, etc.)   10 ft.   3 m
   Total dynamic head   171 ft   52.4 m

On larger systems where conditions vary considerably during the season such as operation of fewer than the maximum laterals, over-extended periods thus should be noted; however, in any case the above maximum conditions will determine the size of the unit or units.

Step 16: Pump and power unit selection

It must be assured that the design engineer has at hand the performance characteristics of pumps and power units. The selection of the pumping unit then becomes a matter of matching the unit to the conditions as previously determined.

Pumps and power unit for the sprinkler system should be analysed from the standpoint of: (a) meeting the operating conditions; (b) comparative fuel costs where there is a choice of power units; and (c) availability of service and repair parts. A competent pump engineer should be consulted if the designer has not previous experience in the field.

The pump engineer should visit the site or have a complete understanding of the following:

1) Is a pumping unit to be in a permanent location or is it to be portable.

2) The maximum suction lift to pump and fluctuation of water level.

3) If water source is from a dug or cased tube-well the well test performance data and physical data on construction.

4) Chemical quality of water and if water contains abrasives or debris that would be harmful to a pump and the system.

5) Altitude and general temperature data during operating seasons.

6) Power sources and fuel costs.

7) Approximate hours of operation per year.

8) Minimum and maximum head capacity requirements and

other data listed under step 15.

9) Provisions in system design for check, discharge and air control valves (pump engineer should list his requirements).

10) Future plan for expansion of system.

The system designer may want some general preliminary data on the power requirements. The following formulas are useful in meeting these calculations.

Metric

$$WHP = \frac{Q \times TDH}{75}$$

$$or \ BHP = \frac{Q \times TDH}{75 \times Eff}$$

where Q = litres/sec pump discharge
TDH = Total dynamic head in metres
Eff = Pump efficiency
In our example

$$WHP = \frac{6.2 \times 52.4}{75}$$

$$= 4.33 \ HP$$

Considering pump efficiency is 60 per cent

$$DHP = \frac{6.2 \times 52.4}{74 \times 0.6}$$

$$= 7.22 \ HP$$

A 7.5 HP motor pumpset will be most suitable.

Cost estimates: Suggested items of components are:

Item	Quantity and unit	Unit cost	Total
Well unit	one		
Pump house			
Power unit			
Pump unit			
Mainline complex			
Lateral line complex			
Sprinklers and risers			
Special equipment			
Installation cost			
Farm units			
a) Labour			

b) Tractor cost
c) Office

Other costs

Total cost of system

The depreciation factors of various equipments/accessories are given below which will help to establish depreciation schedules and estimates of the operation and maintenance costs.

Depreciation factors

Water supply well and casing   20 years

A model sprinkler system design form is given on the following page to help in designing the system for a farm.

SPRINKLER IRRIGATION DESIGN FORM

Owner: Address:

Site Location:

    Village:...............Block:..................Taluk..............

    District:...............State:......................................

....

FARM RESOURCES

1) *Soils*—See soils map which should include:

    Soil types and locations

    Surface texture

    Subsoil texture

    Effective depth of soil (cm)

    Basic intake rates (cm/hr)

    Special problems (*e.g.*, drainage, hardpan, etc.)

    Source of soils information (Dept. of Agriculture, field tests, etc.)

2) *Topography*—See map which should include:

    Contours and contour interval

    Reference points

    Scale

    Wind direction and velocity

    Source of information

3) *Climatic factors*

    Climate zone

    Est. growing season (days)

    Temperatures

    Frost hazard information

    Pan evaporation Annual.........(mm); peak.........(mm/day)

Annual precipitation (cm)

Seasonal precipitation (cm)

Wind (see map)

  Speed and direction

  If variable, then note such

4) *Crops*

Present	Acres	Yield	Future	Acres	Yield
a.					
b.					
c.					
d.					

5) *Water supply*

Source (describe)

Water availability

Water quality (samples and lab report)

Delivery schedule

Seasonal variation in quantity

Screening problems

Pressure available at source

Other water facts (water costs, alternate pumping sites, alternate sources, measurements required, etc.)

Well data—on attached supplement

6) *Desired labour and irrigation operations*

Hours/day ....; Days/week ....

Coverage time (days) ....

Moves/day ....

Labour available ....

Full coverage systems? ....

Labour costs ....

Other owner preferences:

7). *Power availability*

Electrical:

Power phase ....; Voltage ....; H.P. limits.........˙

Internal Combustion:

Full type ....; H.P. limitations....................

Unit costs @ farm ....

CHAPTER 9

Choosing Sprinkler System

Discharge requirements

The required discharge of an individual sprinkler is a function of the water application rate and the two-way spacing of the sprinklers. The formula for computing the required discharge is:

$$Q = \frac{S_1 \times S_m \times i}{360}$$

where Q = required discharge of individual sprinkler lit/sec.

S_1 = spacing of sprinklers along the laterals, metres.

S_m = spacing of laterals along the main, metres.

i = optimum application rate, cm/hr.

Capacity of the sprinkler system

The required capacity of a sprinkler system depends on the size of the area to be irrigated (design area), the gross depth of water applied at each irrigation and the net operating time allowed to apply water to this depth. The capacity of the system may be calculated by the formula:

$$Q = \frac{2780 \, A \cdot d}{F \, H \, E}$$

where Q = discharge capacity of the pump, lit/sec.

A = area to be irrigated in ha.

d = net depth of water application cm.

F = number of days allowed for the completion of one irrigation.

H = number of actual operating hours per day.

81

E = water application efficiency per cent.

In the above equation, it may be noticed that F and H are of major importance in that they have a direct bearing in the capital investment per ha required for equipment. From the formula it is clear that the greater the product of these two factors, the smaller is the capacity for a given area.

Discharge of sprinkler nozzle

The discharge of a sprinkler nozzle may be computed from the following orifice flow formula derived by Toriceli.

$$q = Ca \sqrt{2gh}$$

where q = nozzle discharge, m³/sec.

a = gross sectional area of nozzle or orifice, m².

H = pressure head at the nozzle, metres.

g = acceleration due to gravity, m/sec².

C = coefficient of discharge which is a function of friction and contraction losses.

Water spread area of sprinkler

The area covered by a rotating head sprinkler may be estimated by the following formula suggested by Cavazza

$$R = 1.35 \sqrt{dh}$$

where R = radius of wetted area covered by the sprinkler in metres

d = diameter of nozzles in metres.

h = pressure head at the nozzle in metres.

Maximum coverage is attained by the jet emerges from the sprinkler at an angle of 30°-32° above the horizontal. Most rotating sprinklers are standardised at 30°.

Rate of application

The average rate of application called, precipitation intensity for a single sprinkler may be estimated by the following formula:

$$Ra = \frac{q}{360} \times A$$

where Ra = water application rates, cm/hour.

q = rate of discharge of sprinkler, lit/sec.

A = wetted area of sprinkler, m².

The main consideration in the design of any sprinkler system is the moisture requirements of the crop and the ability of the soil to absorb and retain water. The sprinkler is the key as its operating characteristics under optimum pressures will determine how it will fit into the design. Table 17 gives the various classifications of sprinklers and their adaptability.

Distribution pattern

The typical sprinkler has a geometric pattern of higher application near the sprinkler head and increasing to the outer circumference (Figs. 19, 20). Pressure and wind markedly affect the pattern of any sprinkler. Operating at lower pressure will result in waves in the pattern and large droplets. Higher pressure than recommended also distorts the pattern, usually resulting in a finer 'mist' and marked drifting under wind condition (Figs. 21, 22). It is therefore necessary to follow the pressure recommended for the sprinkler to be selected. Figure 21 gives the methods of assessing sprinkler operating pressure. Further, it is necessary to have a generous overlap of the sprinkler throw.

SPRINKLER OVERLAP

The following are the general guidelines for spacing of the sprinklers:
1) For low to moderate pressure sprinklers: (0.35 to 1 kg/cm²).
   a) For wind conditions up to 6 km/hr do not exceed 50 per cent of the wetted diameter of the sprinkler for the spacing of the sprinkler in the lateral line and 65 per cent of the wetted diameter for the distance of the lateral move to its next position.
   b) For winds 6 to 15 km/hr reduce the above percentage but not to exceed 40 per cent and 50 per cent respectively.
2) For high pressures and giant sprinklers: (4 to 7 kg/cm²).
   It should not exceed 60 per cent of the wetted diameter for both the lateral; and more intervals in winds not to exceed 6 km/hr and reduce to 50 per cent if wind conditions are in the 6 to 15 km/hr range (Fig. 24).
3) Pipe lengths: The pipe lengths for sprinkler systems are standardised. The lateral and portable mainline tubings are available in 3, 6, 9, 12 m lengths. In the interest of economy, the longer pipe lengths having fewer couplers reduce the unit cost.
4) Height of sprinkler: The height is dictated by the crop being

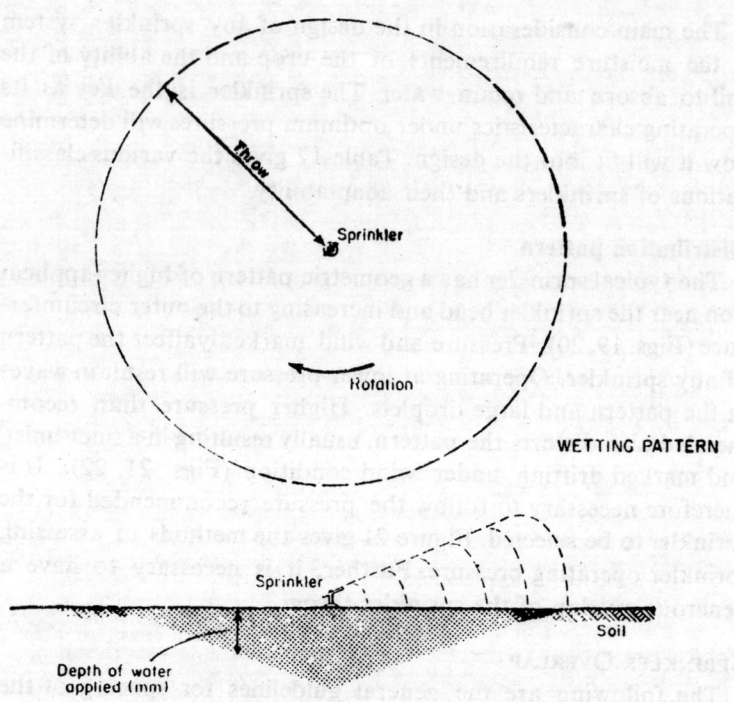

Fig. 19. Distribution pattern—one sprinkler.

irrigated and also to avoid hydraulic turbulance in the riser pipe. Minimum height is as follows for most irrigation sprinklers:

Riser pipe dia. in cm	Minimum riser ht. in cm
1.25	7.5
2.00	15.0
2.50	30.0
7.50	90.0

Most crops exceed 30 cm, so except for clean cultivated orchards where low riser pipes are desirable for under tree sprinkling, the choice will be the minimum height to clear the crop. Although some research studies indicate an additional height from 30 to 60 cm to improve the distribution efficiency. There are obvious disadvantages such as wind drift and handling the lateral lines. Normally farmers prefer 45 to 60 cm risers, except for high growing crops such as cotton and millets.

Table 17. Classification of sprinklers and their adaptability

Type of sprinkler	Low pressure 5-15 p.s.i. 0.35-1 kg/sq. cm.	Moderate pressure 15-30 p.s.i. 1-2 kg/sq. cm.	Intermediate pressure 30-60 p.s.i. 2-4 kg/sq. cm.	High pressure 50-100 p.s.i. 3.5-7 kg/sq. cm.	Hydraulic or giant 80-120 p.s.i. 5.6-8.4 kg/sq. cm.	Undertree low-angle 10-50 p.s.i. 0.7-3.5 kg/sq. cm.	Perforated pipe 4-20 p.s.i. 0.28-1.4 kg/sq. cm.
1	2	3	4	5	6	7	8
General characteristics	Special thrust springs of reaction-type arms.	Usually single-nozzle oscillating or long-arm dual-nozzle design.	Either single or dual nozzle design.	Either single or dual nozzle design.	One large nozzle with smaller supplemental nozzles to fill in pattern gaps. Small nozzle rotates the sprinkler.	Designed to keep stream trajectories below fruit and foilage by lowering the nozzle angle.	Portable irrigation pipe with lines of small preforations in upper third of pipe perimeter.
Range of wetted diameters	20 to 50 feet. 6 to 15 m.	60 to 80 feet. 15 to 24 m.	75 to 120 feet. 23 to 37 m.	110 to 230 feet. 33 to 70 m.	200 to 400 feet. 60 to 120 m.	40 to 90 feet. 12 to 27 m.	Rectangular strips 10 to 50 feet wide. 3 to 15 m.
Recommended minimum application rate	0.40 inch per hour- 1.0 cm/hr.	0.20 inch per hour- 0.50 cm/hr.	0.25 inch per hour- 0.62 cm/hr.	0.50 inch per hour- 1.25 cm/hr.	0.65 inch per hour- 1.6 cm/hr.	0.33 inch per hour- 0.83 cm/hr.	0.50 inch per hour- 1.25 cm/hr.
Jet characteristics (assuming proper pressure-noz-	Waterdrops are large due to low pressure.	Waterdrops are fairly well broken.	Waterdrops are well broken over entire wetted	Waterdrops are well broken over entire wetted	Waterdrops are extremely well broken.	Waterdrops are fairly well broken.	Waterdrops are large due to low pressure.

(Contd.)

1	2	3	4	5	6	7	8
zle size relations).			diameter.	diameter.	diameter.		
Moisture distribution pattern (assuming proper spacing and pressure-nozzle size relations).	Fair	Fair to good at upper limits of pressure range.	Very good.	Good except where wind velocities exceed 4 miles per hour.	Acceptable in calm air. Severely distorted by wind.	Fairly good, Diamond pattern recommended where laterals are spaced more than one tree interspace.	Good pattern is rectangular.
Adaptations and limitations.	Small acreages. Confined to soils with intake rates exceeding 0.50 inch per hour and to good ground cover on medium- to coarse-textured soils.	Primarily for undertree sprinkling in orchards. Can be used for field crops and vegetables.	For all field crops and most irrigable soils. Well adapted to overtree sprinkling in orchards and groves and to tobacco shades.	Same as for intermediate pressure sprinklers except where wind is excessive.	Adaptable to close-growing crops that provide a good ground cover. For rapid coverage and for odd-shaped areas. Limited to soils with high intake rates.	For all orchards or citrus groves. In orchards where wind will distort overtree sprinkler patterns. In orchards where available pressure is not sufficient for optimal operation of overtree sprinklers.	For low-growing crops only. Unsuitable for tall crops. Limited to soils with relatively high intake rates. Best adapted to small acreages of high-value crops. Low-operating pressure permits use of gravity or municipal supply.

Source: Sprinkler Irrigation Guidebook, USAID, Washington, 1968.

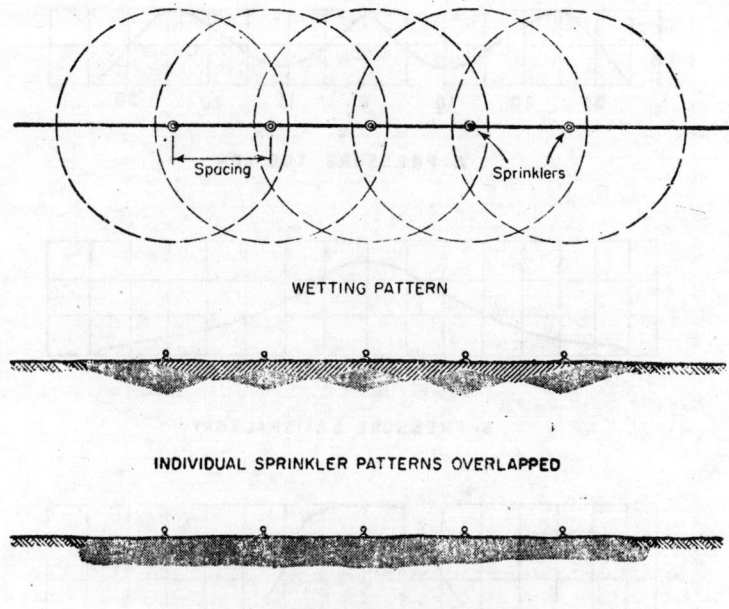

WETTING PATTERN

INDIVIDUAL SPRINKLER PATTERNS OVERLAPPED

RESULTING DISTRIBUTION OF WATER IN SOIL

Fig. 20. Wetting and distribution patterns from several sprinkler operating close together.

EVAPORATION LOSSES

Evaporation losses are made up of sprinkler spray evaporation and surface evaporation from the free water on the plant and soil. These losses are low during sprinkling. A typical farm system operating at 2 to 3.5 kg/cm^2 has negligible losses under cool weather and low wind conditions. Even at high wind conditions at 25 km/hr, the losses for 24 hour sprinkling period on a growing crop do not exceed 8 per cent. On bare ground the losses may be double that amount. Therefore 24 hours a day operation is feasible for irrigating the crop without much loss.

Many other factors should be considered when deciding which is the best sprinkler system to use. These include land topography, field shape, soils, crops and labour.

LAND TOPOGRAPHY

Sprinkler can be used on hilly and uneven land which is unsuitable for surface irrigation. The type depends on the land slope.

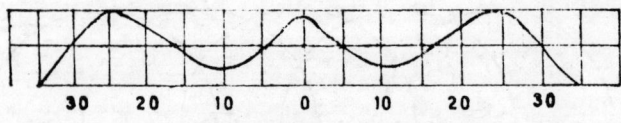

A-PRESSURE TOO LOW

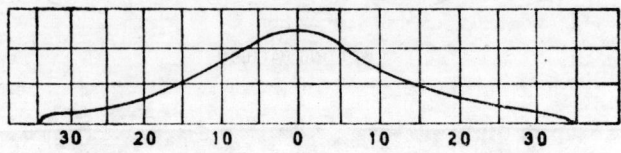

B-PRESSURE SATISFACTORY

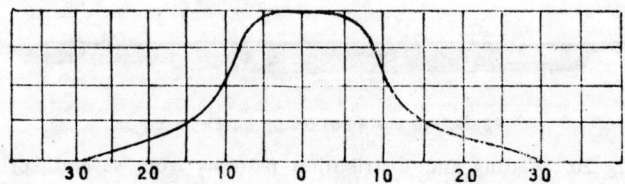

C-PRESSURE TOO HIGH

Fig. 21. Effect of different pressures.

If the land slope is less than 5 per cent any type can be used. On steeper slopes, it becomes difficult to keep mobile rainguns in line when moving across the slope. Slopes greater than 15 per cent are only suitable for conventional, portable, permanent and semi-permanent systems. The soil erosion should be avoided. Land which is undulating and irregular can present problems for most systems.

FIELD SHAPE

All the systems are easily adapted to regular shaped fields such as squares or rectangulars. Conventional mobile rainguns and small side move systems can be adapted to irregularly shaped fields. Centre pivots only irrigate circular areas and as much as 20-25 per cent of a square field will be lost unless a special corner irrigating device is available.

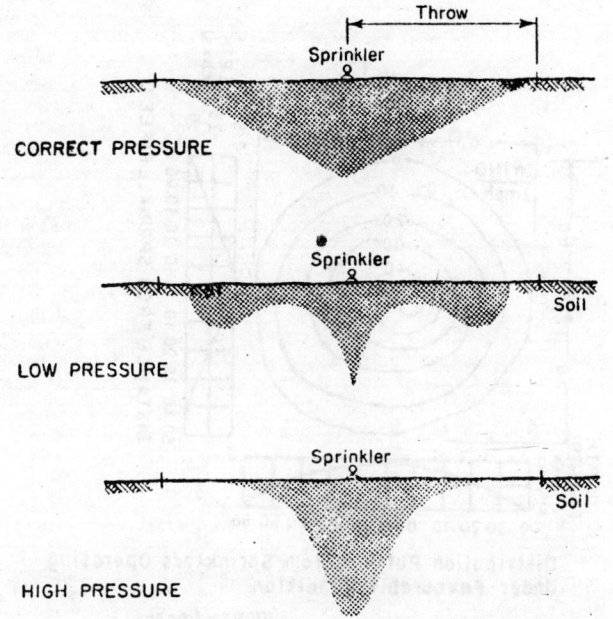

Fig. 22. Effect of operating pressure in sprinklers.

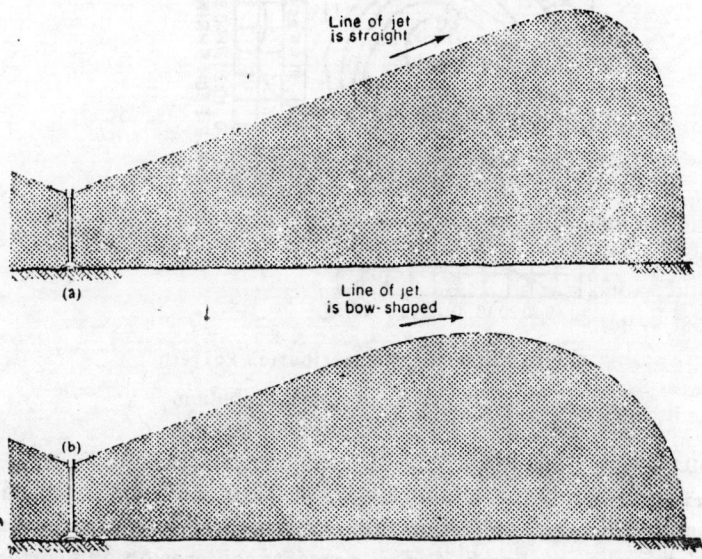

Fig. 23. Assessing sprinkler operating pressures: (a) correct, (b) too low.

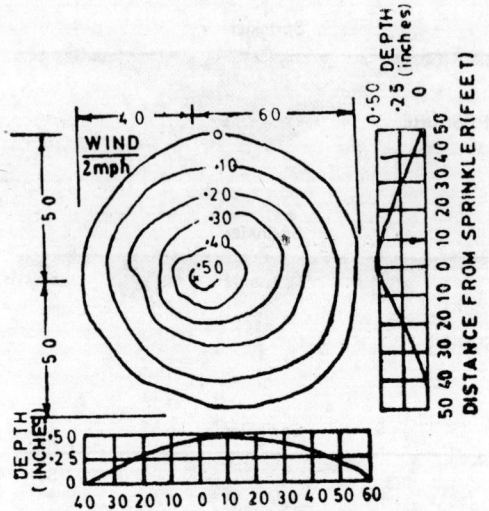

Distribution Pattern From Sprinklers Operating
Under Favourable Condition

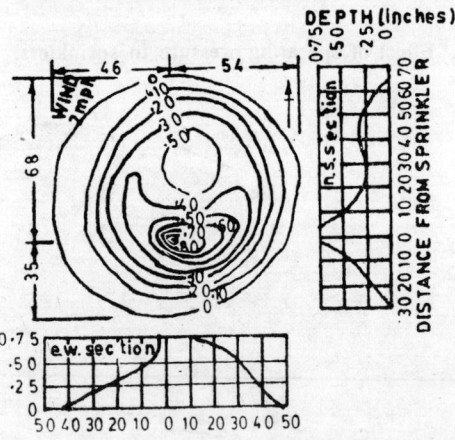

Effect Of Wind On Distribution Pattern

Fig. 24. Effect of wind distribution.

SOILS

A sprinkler system should always be adopted for the soil conditions so that the application rate is always less than the rate at which soil can absorb it. This prevents soil erosion and surface runoff. Since in clay soil, the intake rate is slow, the systems should give less than the absorption rate. If the soil surface structure is easily damaged under irrigation, then mobile rainguns should not be used.

CROPS

Most crops can be irrigated using all systems. Mobile rainguns may cause damage to young plants and delicate crops such as tomatoes.

LABOUR

Conventional portable systems require large labour groups. Other systems are designed to reduce labour needs to a minimum. This includes conventional permanent and semi-permanent systems, mobile rainguns, and mobile lateral systems. Many of these rely on machinery and so the few men that are required must be highly skilled in operation and maintenance.

CHAPTER 10

Special Use of Sprinkler Equipment

The introduction of light weight aluminium tubing and quick action couplers has led to several companion uses of sprinkler equipment on the farm. Although all sprinkler systems are designed primarily for meeting the water requirement of crops, most systems have uses, which because of economic benefits may justify large investments in the equipment. Ina lmost all climatic conditions, these potential benefits should be considered and incorporated whenever possible in the design and management of the system. Generally, a farm sprinkler system is designed for a specific area and cannot be spared for other applications, unless they occur during the off irrigation season. The other more important uses of the sprinkler systems are discussed in detail.

a) SEED GERMINATION

The uniformity of crop stand and time of maturity is of economic importance for many crops, particularly for vegetables. The sprinkler system often ensures adequate seed germination with only one light application of water after seeding. Such applications are more efficient than surface irrigation methods.

b) APPLICATION OF FERTILISERS

Dissolving soluble fertilisers in water and applying the solution through a sprinkler system is quick, economical, easy and effective. A minimum of equipment is required and once the apparatus for adding the fertiliser to the irrigation water is set up the

92

crop being irrigated can be fertilised with less effect than is required for mechanical application. Penetration of fertiliser into the soil can be regulated by the time of application in relation to the total irrigation period. The fertiliser can be dissolved in water in a barrel or a closed container. There are several advantages in using sprinkler irrigation systems as a means of distributing fertilisers. First, both irrigation and fertilisation can be accomplished with a little more labour than is required for irrigation alone. Second, close control can usually be maintained over the placement depth of fertiliser as well as over lateral distribution. The uniformity of distribution can be only as good as the uniformity of water distribution.

The simplest method of applying fertiliser through a sprinkler system is to introduce suction into the system at the suction side of a centrifugal pump (Fig. 25). A pipe or hose is run from a point near the bottom of the fertiliser solution container to the suction pipe of the pump. A shut-off valve is placed in this line for regulating pumps. Another pipe or hose from the discharge side of the pump to the fertiliser container provides an easy method of filling the container and dissolving the fertiliser and for rinsing. If a closed pressure type container is used, the lines from the discharge side of the pump can be left open and the entrance of the solution into the water can he regulated by the valve into the suction side of the line.

The common procedure followed in applying fertiliser through sprinkler systems consists of three time intervals. During the first interval, the system operates normally, wetting the foliage and the soil. In the second interval, fertiliser is injected into the system. This application should rarely be for less than 30 minutes and preferably for an hour and a half. This eliminates the possibility of poor distribution due to slow or uneven rotation of sprinklers. Also with normal fertiliser application rates, the solution passing through this system will be more diluted. This lessens the possibility of foliage burn or damage to the system corrosion. The third time interval should be long enough to completely rinse the system with clear water and removing fertiliser from plant foliage. Depending on the rate of application at which the system is operating this last rinse off should be between 30 minute intervals for fast rates and one and a half hours for slower rates. The last time interval also has the effect of

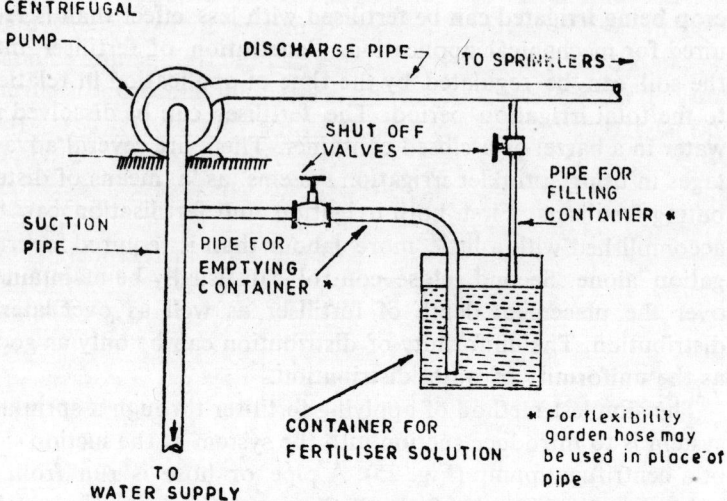

Fig. 25. Method for adding fertilisers in solution to centrifugal
pump system.

moving the fertiliser down into the crop root zone.

c) SOIL AMENDMENTS

Various soluble soil amendments, such as gypsum, sulphuric
acid, limes, and soluble resins can be applied through sprinkler
systems. The method used is the same as those used to add soluble fertilisers.

d) FROST PROTECTION

Sprinkler irrigation systems are being used for frost protection. The ordinary system is limited because of the area it can
cover with any one setting of the lateral lines. Therefore, it is
necessary to add more lateral lines and sprinklers at predeter-
mined spacing so that the entire field can be covered with a
fine mist of water during freezing temperature. This has been
used successfully on low growing and other vegetable crops, such
as, tomato.

e) APPLYING INSECTICIDES AND WEED CONTROL CHEMICALS

Attempts have been made to inject insecticides, fungicides
and weed control chemicals through the sprinkler system in a
way similar to the fertiliser application.

f) COOLING CROPS

Many crop yields are seriously depressed by excessively high air temperature during the fruiting period. Temperature above 35°C may cause blossom drop of beans and fruit drop of citrus, and temperatures in excess of 38°C for several days can cause losses of grapes up to 50 per cent. Sprinkler systems which apply time sprays with low application rates will reduce ambient air temperature and leaf temperature up to 5°C or more. Crop losses can thus be minimised and fruit quality maintained. Cooling crops by sprinkler requires a full coverage system, but unlike frost protection, water can de applied intermittently (15 minutes off, 15 minutes on), thus conserving water. This method also requires water of fairly good quality.

g) LINE SOURCE IRRIGATION

To bring more information about water use, fertiliser and water interaction for many crops at short duration, line source irrigation method is used. In this method, a sprinkler lateral is used for irrigation for the various crops in the field. Since water distribution is more near the sprinkler rise; and reduced uniformly if you go away from the line. The amount of water used (irrigated) in different zones and finally the yield of the crop can be obtained. This will be useful to decide the amount of water required to get maximum production per unit quantity and also fertiliser and water production functions.

h) OTHER USES

There are numerous other uses for sprinkler irrigation equipment, both on the farm and elsewhere. The following are some of them:

   i) Cooling livestock and poultry environments.
   ii) Farm fire protection.
   iii) Water distribution for compaction of earth fills.
   iv) Settling of dust.
   v) Log curing.

CHAPTER 11

Pumping Plant

A pump is a machine which changes mechanical energy produced by an internal combustion engine or electric motor into useful water energy. In sprinkler irrigation this energy provides the pressure and discharge needed to distribute water in the mainline and laterals to the sprinklers.

Centrifugal pump

There are several different types of pumps available to meet the needs of a wide range of tasks. The most common type used in sprinkler irrigation is the centrifugal pump. It is best suited to the pressure and discharge requirements of sprinklers as basically it is simple in design, easy to use and relatively inexpensive to buy and maintain. They are simple in construction, easy to operate, low in initial cost and produce a constant steady discharge. The wearing parts are few. They are adapted directly to motor or engine drives without the use of expensive gears.

Principles of operation of centrifugal pumps

The centrifugal pump operates on the principle of centrifugal action. In a centrifugal pump, a motor or other drives rotate an impeller fitted with vanes immersed in water and enclosed in a casing. Water enters the case at the centre and is immediately engaged by the impeller which is in rapid rotation. This rotation causes a flow from the centre of the impeller to its rim or the outside of the case when a pressure head is rapidly built up. To relieve this pressure, the water escapes through the discharge pipe. The centrifugal pump will not operate until the case is entirely full of water or primed. The need of priming is one of the

96

disadvantages of the horizontal centrifugal pump.

Centrifugal pumps are built on two types—the horizontal centrifugal and the vertical centrifugal. The horizontal type has a vertical impeller connected to a horizontal shaft. The vertical centrifugal pump has a horizontal impeller connected to a vertical shaft. Both types of centrifugal pumps draw water into their impellers, and therefore they must be set only a relatively few metres above the water surface. The centrifugal pump is limited to pumping from reservoirs, lakes, ponds, streams and shallow wells where the total suction lift is not more than eight metres.

The horizontal centrifugal pump is the most commonly used in irrigation and the sprinkler system. To keep the suction lift within operating limits, the horizontal type can be installed in a pit but it usually is not feasible to construct watertight pits more than about three to four metres deep (Fig. 26).

Characteristics

The principal characteristics of a centrifugal pump are:

a) Smooth even flow—easy on pump, motor, piping and foundation.

b) adapted to high speed operations and to different speeds.

c) Nonoverloading of power unit with increased heads but there may be some danger of overloading if head is decreased.

d) Capacity and head depend upon r.p.m. and impeller diameter and width. In a given pump, the capacity and head will vary according to the individual operating characteristics of that pump; that is, an increase in head reduces the capacity and vice versa.

e) Horse-power is a function of capacity, head, and pump efficiency.

f) When the speed is kept constant, capacity decreases as head increases and power is reduced. Likewise, when the head is reduced, capacity increases and power goes up.

g) When the operating speed is changed (Fig. 27) the capacity will change in direct proportion to the variation in speed. At the same time, the head will vary as a square of the change in speed while horse-power will change as the cube of the change in speed. This is represented by the following formula (variable speed-diameter constant):

$$\frac{\text{r.p.m.}_1}{\text{r.p.m.}_2} = \frac{\text{cap.}_1}{\text{cap.}_2} = \frac{\sqrt{\text{head}_1}}{\sqrt{\text{head}_2}} = \frac{\sqrt[3]{\text{hp}_1}}{\sqrt[3]{\text{hp}_2}}$$

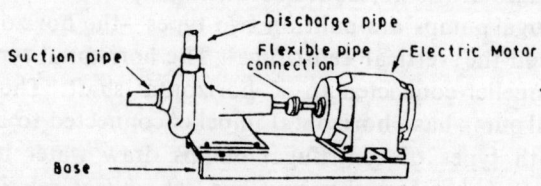

DIRECT CONNECTED HORIZONTAL CENTRIFUGAL PUMP
(Internal combustion on electric motor may be used)

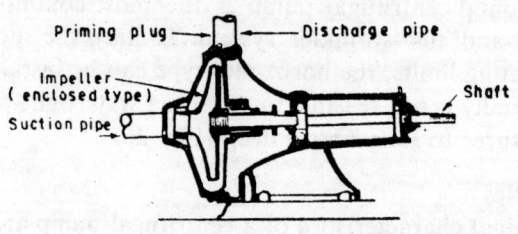

CROSS SECTION OF MODERN HORIZONTAL CENTRIFUGAL
PUMP
(single suction enclosed impeller)

Fig. 26. Horizontal centrifugal pump for surface or pit installation.

h) When it is necessary to vary the characteristics of a pump operating at constant speed, the same relationships expressed in (f) above hold except that here it is the diameter of the impeller that is changed. Then the capacity varies directly with the diameter; the head varies as a square of the diameter; and the horsepower varies as a cube of the diameter. This is expressed by the following formula (variable diameter-constant speed):

$$\frac{dia._1}{dia._2} = \frac{cap._1}{cap._2} = \frac{\sqrt{head_1}}{\sqrt{head_2}} = \frac{\sqrt[3]{hp_1}}{\sqrt[3]{hp_2}}$$

i) These changes (g and h) take place with little or no change in efficiency for small changes in speed and impeller diameter (maximum increase of speeds of about 5 per cent). For large changes in speed or impeller diameter, the efficiency will be reduced.

CHARACTERISTIC CURVES

For a particular job, the best selection of a pump will be one

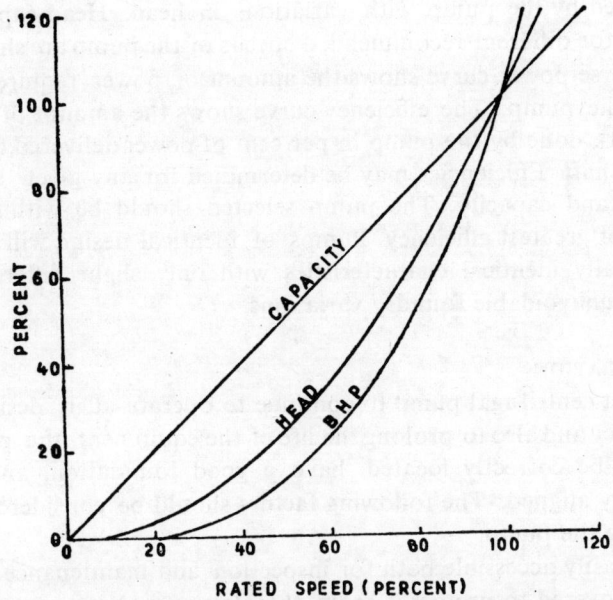

Fig. 27. Effect of speed change in centrifugal pump performances.

that will operate at its peak efficiency. Unfortunately, this is rarely possible for there is only one capacity and one head condition for each pump where the highest efficiency is obtained. Because it is obviously impossible for any manufacturer to design and build the many pumps required to meet all operating conditions, manufacturers have settled upon standard designs for required head and capacity ranges. A will-designed and integrated line of pumps will be so arranged that it is possible to select some pump from the line for any condition and obtain an efficiency that is within a few percentage points of the maximum. Characteristic curves are available and should be used to select the best pump for the particular job.

These curves have been developed at the factory after exhaustive tests during which the water capacity, pressure, power input, etc., are carefully measured and plotted on a curve.

A full set of characteristic curves includes, in addition to the head-capacity curve for different speeds, an efficiency curve and a horse-power curve. The head-capacity curve for the constant speed of the pump represents the varying quantities of water delivered by the pump with variations in head. Head-capacity curves for different recommended speeds of the pump are shown. The horse-power curve shows the amount of power required to drive the pump. The efficiency curve shows the amount of usable work done by the pump in per cent of power delivered to the pump shaft. Efficiencies may be determined for any given head, speed, and capacity. The pump selected should be within the range of greatest efficiency. Pumps of identical design will have practically identical characteristics with only slight differences due to unavoidable foundry variations.

INSTALLATION

For a centrifugal pump to continue to operate at its designed efficiency and also to prolong the life of the equipment, the pump should be correctly located, have a good foundation, and be properly aligned. The following factors should be considered in locating the pump:

1) Easily accessible both for inspection and maintenance.

2) Covered to protect it from the elements. A house can be used on permanent installations. In the case of a house, available headroom should be provided for servicing the equipment.

3) Safeguarded against flood conditions unless a wet pit-type pump is used.

4) Placed as close as possible to the water supply so as to make the suction line short and direct.

Pumping units that are to be installed in a permanent location provide the opportunity for developing the best type of foundation (Fig. 28). Concrete is the best material for constructing a good pump foundation. The pump unit should be securely fastened to the foundation. A recommended method of setting the foundation bolt is shown. The coupling between the pump and power unit must be in correct alignment regardless of the type of coupling. Figure 28 shows how a coupling can be checked for alignment with a steel straight edge. When the coupling is by a shaft and double universal joint, a shield should be placed over and around the two horizontal sides to protect the operator from the fast moving shaft.

To operate properly, the pump must be at a level position at all times. Figure 28 shows how four to six wedges can be used to raise the entire pumping unit about 2 cm above the foundation. The wedges can then be adjusted as necessary to bring the pump into a level position. After the pump has been levelled, a dam should be built at least 6 cm high around the base plate; then concrete poured in, to the required depth and allowed to harden thoroughly. The wedges may be left in place. When the concrete is hardened, the foundation bolt should be tightened and a recheck made of the alignment. If there is any misalignment, it can be corrected by placing shims under the pump, motor, or brackets.

The motor should now be checked to see that it rotates in the proper direction. The rotation of the motor must be in the same direction as the arrows on the pump casing. It is important that pumps line up naturally with their power unit and piping. Pipes

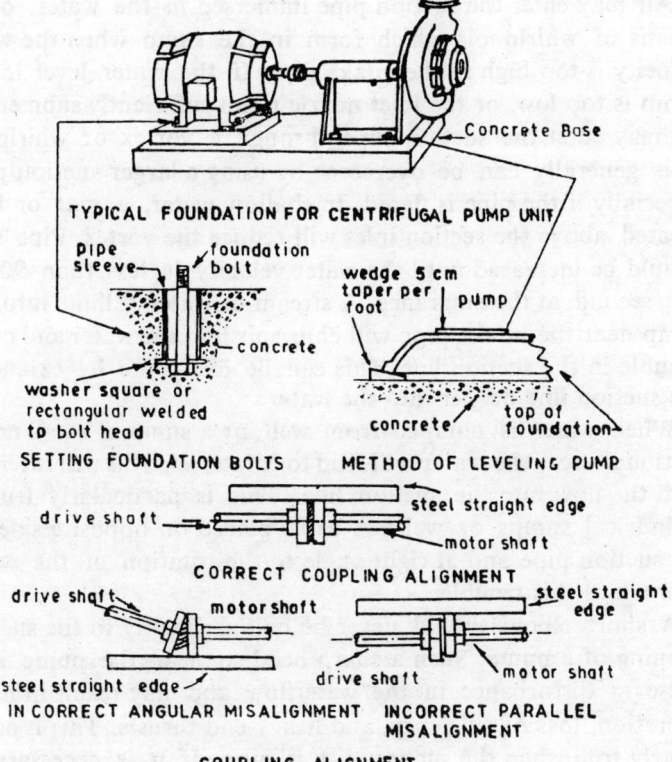

Fig. 28. Installing centrifugal pump.

should not be forced into place with flange bolts as this may draw the pump out of alignment. Suction and discharge pipe lines should be supported independently of the pump so as not to put any strain on the pump casing.

The suction pipe, particularly in the case of long intake pipes and high suction lifts, should be laid with a uniform slope, upward from the source of water to pump. There should be no high spots where air can collect and cause the pump to lose its prime. The inlet end of the suction pipe should be suspended above the earth bottom of a stream or pond or laid in a sump made of concrete or metal. On horizontal suction lines where a reducer is used, it should be of the eccentric type with the straight section on the upper side of the line and the tapered section on the bottom side.

Air may enter the suction pipe immersed in the water, or by means of whirlpools which form in the sump when the water velocity is too high in the intake pipe. If the water level in the sump is too low, or the inlet nozzle is not sufficiently submerged, air may enter the suction pipe through a vortex or whirlpool. This generally can be overcome by using a larger suction pipe, especially if the pipe is flared. In shallow water, a mat or float located above the section inlet will reduce the vortex. Pipe sizes should be increased until the water velocity is less than 90 cm per second at the enterance. A stream of water falling into the sump near the intake pipe will churn air into the water and cause trouble in the suction line. This can be overcome by extending the suction line deeper into the water.

When water is pumped from well or a sump of small cross-sectional area, the water will tend to rotate, and this will interfere with the flow into the suction line. This is particularly true in cylindrical sumps or wells. A baffle placed on opposite sides of the suction pipe and at right angle to the rotation of the water overcomes this trouble.

A short elbow should never be bolted directly to the suction opening of a pump. Such a sharp bend so near the pump inlet causes a disturbance in the waterflow and may result in noisy operation, loss of efficiency, and heavy end thrusts. This is particularly true when the suction lift is high. If it is necessary to make a bend in the suction line, it should be in the form of a long sweep or long radius elbow and should be placed as far away from the pump as is practicable.

Screens or strainers should be used to exclude debris from the suction line. If the source of water contains large amounts of small debris, a screen placed around 60 or 90 cm from the inlet of the suction hose will provide good protection and be less likely to clog. Strainers are generally small and are fastened to the end of the suction pipe. They are satisfactory in relatively clear water.

In some cases, it is not possible to locate the centrifugal irrigation pump in a permanent location. It may be needed in more than one location on the farm. This increases the difficulty of providing a proper foundation. Portable pump units generally are mounted on wheels or skids. It is highly important to locate this type of unit so that it is level, is on firm ground, and is securely staked in place and will not shift during the time it is operating.

In pumping from rivers with moderately sloping banks, the horizontal centrifugal pump may be mounted on skids, on sloping timbers or track so that it can be removed quickly from floods. This method also can be used where the water level fluctuates sufficiently to be out of range of suction lift if the pump were installed in a permanent location. With steep banks it may be necessary to build a foundation platform secured to piling or to place the pump unit on a floating barge or boat.

PRIMING

Centrifugal pumps, due to their non-positive action, must be primed before the pump operates. They will not lift water from a source of supply unless the pump casing and the suction pipe are both full of water. This can be accomplished by one of the following priming methods that are generally used in irrigation pumping:

1) By use of a foot valve and water from an outside supply. The outside supply must be large enough to keep the pump and suction line filled until the pump is primed. To prime, close discharge gate valve, open air vent valve, and open gate valve in supply line until all air is expelled and water issues from vent openings. Close valve in supply line, close air vent valves, and start pump; then open discharge gate valve.

2) By separate hand-controlled priming pump and foot valve.

The hand-priming pump is a simple, high-speed air pump with its primer suction inlet connected to the priming part of the centrifugal pump. If connection is made on the pump discharge, a valve must be installed in the priming line. The pump handle is used to actuate a diaphragm in the priming-pump chamber. Air is drawn into the chamber from the centrifugal pump through a suction valve on the 'upstroke' and discharged through a discharge valve on the 'downstroke.' To prime the pump, close the discharge gate valve and air vent valve. Open valve in priming line. Exhaust air from pump and suction piping until water flows from priming pump. Close valve in priming line, start centrifugal pump, and open discharge gate valve.

3) Self-priming centrifugal pumps. Self-priming centrifugal pumps are made by several manufacturers. With this type of pump the pump chamber and hopper must be first filled with water. Therefore, its advantage is primarily confined to the smaller size pump. They are used extensively by contractors but are generally limited to small irrigation systems.

After the pump is filled with water, the engine is started, and the water within the impeller is discharged upward into the chamber. This action instantly creates a vacuum at the impeller eye. Air from the suction line and water within the pump rush into this void. They are mixed at the impeller periphery and discharged upward into the chamber where the air escapes from the water. This force of gravity pulls the heavier air-free water down to the impeller. More air is entrained and the cycle is repeated until the pump is primed.

When the pump is primed the water is no longer circulating within the pump while pumping. The pump is equipped with a check valve at the suction inlet to the pump, and thus the pump is always full of water and priming is automatic after the pump is once filled by hand.

TROUBLE CHECKLIST

When the centrifugal pump fails to operate or discharge or pressure drop, the cause of trouble should be investigated immediately and steps taken to eliminate it. Investigation shows that the majority of troubles with centrifugal pumps, except mechanical failures, can be traced to the suction line, its joints, elbows, foot valves, and other accessories. Air leaks in the suction line must be eliminated to attain the maximum suction lift for a given installation. The following checklist will be helpful in locating the cause of the trouble:

Pump fails to prime

1) Failure of the pump to prime is mostly occasioned by an air leak in the suction line or in the pump.

2) The most common sources of air leaks are in the threaded connection of the suction line. Coat these connections with pipe cement or white lead and then draw them tight. All connections provided with gaskets must be drawn up tight.

3) The check valve on the discharge side of the pump may have debris lodged between the rubber flap and the valve seat. This will prevent the valve from sealing and forming an air-tight joint.

4) Occasionally, gaskets shrink and admit air into the pump. Tightening the flanges or connections will remedy this difficulty.

5) Rotary shaft seals may leak air if improperly greased or worn. Check this by running the pump and squirting oil on the shaft just outside the seal. If oil is drawn into the seal, a leak is indicated. Filling the seal with grease may eliminate the difficulty, but if the parts are worn, repairs may be necessary. If the seal is always kept full of the proper grade of grease, little, if any trouble will be encountered.

6) Connections in the priming line between the pump and primer must be air-tight or the pump will fail to prime.

7) Screw tight all drain and fill plugs in the pump case to prevent air leaks.

8) A plugged suction line or a collapsed suction hose liner is a frequent source of priming difficulties. Do not overlook this possibility.

Pump fails to develop sufficient pressure or capacity

1) Check pump speed. The capacity of the pump will vary directly with speed, and pressure will vary with the square of the speed. This means that increasing the speed by 20 per cent will increase the capacity 20 per cent and the head by 44 per cent. On internal engines, check the governor and adjust if necessary. With electric motors, check to see if the motor is across line, wiring correct and receiving full voltage.

2) Check the suction line, strainer, and foot valve. They may be clogged with debris. A frequent source of difficulty is a collapsed suction-hose liner which has the effect of reducing the capacity and pressure the pump develops. The foot valve may be too small or not immersed deep enough to prevent air being drawn in with the water.

3) Check for air leaks in pump or suction line. Air leaks in the suction line or in the pump will occasion a reduction in both capacity and pressure. A small air leak which is not great enough to prevent the pump from priming may reduce both capacity and pressure.

4) Check suction lift. If the suction lift is too high, reduction in capacity will occur. Lifts of more than 6 metres are definitely too high for efficient operation, and the closer the pump can be located to the source of supply, the better will be the results obtained. Refer to Determining Operating Conditions, and discussion on computing suction lift.

5) Check length of suction lines. Long suction lines have the same effect as a high suction lift because of the increased friction when the water passes through the line.

6) Check for worn parts. Worn parts, such as impeller wear rings, will reduce both capacity and pressure. The impeller may be damaged or the casing packing defective.

7) Check impeller for clogging. If the impeller is plugged with foreign material, a reduction in both capacity and pressure will occur.

8) Check piping layout. It is characteristic of centrifugal pumps operated at constant speed that as the pressure is increased, the capacity decreases. In those cases where the pump pressure and capacity are in accordance with the characteristic curve, and when the speed of the engine cannot be increased, if necessary make some alterations in the pipe line so as to reduce the frictional resistance and thereby increase the capacity of the pump.

Pump takes too much power

1) Check speed of pump. If it is higher than rating, reduce speed to pump rating.

2) Head may be lower than pump rating, thereby pumping too much water.

3) Check for mechanical defects such as bent shaft, binding

rotating elements, too tight stuffing box, or misalignment of pump and driving unit.

Pump leak excessively at stuffing box

1) The packing may be worn out or not properly lubricated.

2) The packing may be incorrectly inserted or not properly run in.

3) Packing is not the right kind or the shaft may be scored.

Pump is noisy

1) Hydraulic noise—cavitation-suction lift is too high.

2) Check for mechanical defects such as bent shaft, binding rotating parts, loose, broken or worn out bearings or misalignment of pump and driving unit.

FACTORS IN SELECTING THE SIZE OF PUMPING PLANT

The main factors to be considered in selecting the correct size of the pumps are: (a) crop and cropping pattern including the extent of area irrigated; (b) water requirement including peak water use for these crops; (c) availability of water in the well; (d) availability of funds for investment by the farmer; and (e) availability of the desired equipment in the market.

At present the selection of the pumpset by the farmers is done without taking into account the above factors, but it is arbitrarily decided. In many cases, the capacity of the pumps is far greater than the requirement and hence a huge amount is invested and energy is also wasted. The two important factors to be considered are the availability of water in the well and the peak quantity of water to be pumped. If the water is plentiful in the well and the availability is not a constraint, the selection of the pump will depend on the peak water use of the crops taking into account the various crops which are irrigated from the well. If water is the limiting factor, the selection depends upon the maximum available water in the well. In both cases if proper pumping plants are not selected, considering the demands, then the amount invested will not be economical. Availability of a pumpset is not a problem at present since many manufacturers make varied types of pumpsets. The peak consumptive use can be worked out based on the crops grown in the command. The discharge of the pump required can be computed from the cumulative water requirements of the crops grown and the operating period in hours per day, the efficiency of the pumpset and motor and the efficiency of conveyance and distribution of water. Once the above details are known, the selection of the pumpset is to be done based on the pump characteristics curve given by the manufacturers to suit the requirement of the farmer. Each make has its own specifications and characteristics. The graphs are drawn to give information about discharge versus head, discharge versus efficiency and discharge versus horse-power of the pump. Knowing the total head to be lifted for a particular discharge, the best combination for giving high efficiency and lower horse-power pump can be selected. Though two farmers may have the same acreage and cropping pattern in the same place, if the water table position in their wells are different, the horse-powers of the pump selected for their wells may be different. Similarly, in places where the water table may be common, but the cropping pattern and the area to be irrigated are varying, the horse-power requirements may differ immensely. In some areas, when the soils of two farmers are different, the capacity and horse-power of the pumpset may also vary for similar situations. But all these factors are not considered by the farmers while selecting the pumpsets. They simply get a pumpset as in use by their neighbour without consideration of other factors.

In recent years, one or more pumpsets are installed to irrigate the same area which was covered by one pumpset. This is because after partitioning the property, each one wants to have his own well and pumpset and does not want to share with his brothers. In this process, he provides the same capacity pumpset though the cropped area is reduced by half or one-third. This affects not only the load on the grid by operating the additional pumpsets simultaneously but also involves unnecessary expenditure to install more pumpsets which are not really required. There is no agency to guide the farmer properly. The manufacturer and his agent are interested in selling their produce. The electricity authorities are concerned only for providing electric connections after the pumpsets are installed by the farmer. The extension officers are anxious that more wells are energised so that more areas can be brought under irrigation in order to increase the production. Hence no one is bothered about the optimum utilisation of the resources and energy.

CHAPTER 12

Operation and Maintenance of the Sprinkler System

The supplier of sprinkler irrigation equipment is expected to instal the system properly and to provide the user with instructions for proper operation and maintenance. The extension officer in the field should also provide the necessary on-site assistance as may be required to attain the objectives of conservation irrigation. The farmer should be given instructions on the layout of mainlines and laterals, the spacing of sprinklers, the movement of lateral lines, the time of lateral operation and the maintenance of design operating pressures. He should also be shown how to estimate soil moisture conditions in order to determine when irrigation is needed and how much water should be applied.

A sprinkler irrigation system, like other farm equipment needs maintenance to keep it operating at peak efficiency. Parts of the system subject to the most wear are the rotating sprinkler heads, the pump infilter and the power unit.

Examine the sprinkler heads after each season's use. Replace worn, bent or damaged parts. Any wear on the pump infilter reduces pump efficiency, which increases power costs. The power unit should also be examined for wear and placed in top operating condition before the start of each irrigation season. Preventive maintenance saves money by correcting causes of system failures before they actually occur. Store portable aluminium pipes, couplers and sprinklers in a dry place when they are not in use. Proper storage extends the life of the system parts (Fig. 29).

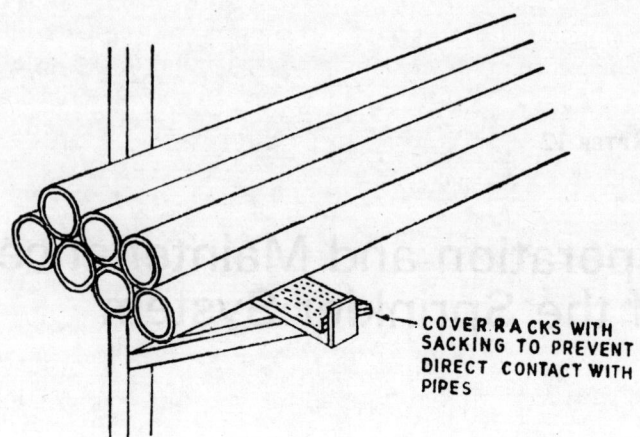

COVER RACKS WITH
SACKING TO PREVENT
DIRECT CONTACT WITH
PIPES

Fig. 29. Store pipes when not in use.

Protect against debris getting into the main and lateral. This is important as a 'plugged' nozzle is not only aggravating but can materially effect application efficiency. This can be accomplished by: (a) screening at the suction inlet to remove any foreign material that would plug the smallest nozzle diameter; (b) starting the system and first flushing out the mainline prior to opening the lateral valve; and (c) flushing out the lateral on every move. Lateral flushing can be done by having a reduced flow in the lateral as it is being connected.

Proper design of a sprinkler system does not in itself ensure success. The system should be operated in keeping with good irrigation practice. It should be ensured that the prime mover and the pump are in alignment, particularly in the case of tractor driven pumps. Service and installation proceeds in respect of the pump and power units should be strictly observed.

While laying the main and lateral pipes, always begin by installing the pump. This necessarily gives the correct connection to all quick coupling pipes and while joining couplings, it ensures that both the coupling and the rubber seal rings are clear.

In starting the sprinkler system, the motor or engine is started with the valves closed. The pump must attain the pressure stated on the name plate, or otherwise there is a fault in the suction line. After the pump reaches the regulation pressure, the delivery valve is opened slowly. Similarly, the delivery valve is closed after stopping the power unit. The pipes and sprinkler lines are shifted as required after stopping. Dismantling of the installation takes place in the reverse order of the assembly described above.

In the event of the sprinkler not operating satisfactorily the following points should be checked:

a) PRESSURE

The sprinkler may not operate or operate poorly due to insufficient or excess pressure. In the case of insufficient pressure, the jet emerges from the nozzle with little or no break up. The sprinkler may turn very slowly, distributing the water in low distinct circles. Check the pressure with a pitot tube (gauge). Low pressure may be a result of overloading the mainline or the pump. If so, close a lateral or a number of laterals until a satisfactory operating pressure is reached. In case of a single lateral operating at low pressure, check for leaks in the lateral or for overloading of the lateral and the many sprinklers mounted on the line.

Excess pressure is generally visible to the eye from the extreme fragmentation of the jet, resulting in a fine mist, invariably drifting away even with the slightest wind. In some cases, the sprinkler may be seen to turn in an anti-clockwise direction. Check the pressure with the pitot guage. Reduce the pressure by partially closing the riser valve (some valves are specially equipped with a counter-locking device). Other remedies are to regulate the mainline valves or to open more laterals.

b) BLOCKAGES

Sprinkler laterals should be flushed before operation in order to minimise blockages. Flushing may be carried out by one of the following methods:

i) *Use of the end flush valve*

On assembly of the line leave the flush valve open. Open the riser valve after assembly is complete and walk to the end of the line. When the water flowing from the end of the lateral is free of soil, straw, grass etc. close the valve.

ii) *Without the flush valve*

While assembling the line leave the riser valve slightly open.

This will enable the flushing of each pipe as assembled. On completing the assembly insert the end plug, walk back to the riser
and open the valve. This method must be used with caution and
never when a large number of laterals are being simultaneously
assembled—as this will cause an unnecessary load on the water
source and resultant drop in pressure. All blockages during irrigation should be immediately attended to.

Blockages may be removed with a piece of steel wire. It may
be necessary to remove the nozzle in order to expel a blockage.
This may be done while the lateral is still operating and will ensure
good flushing of the sprinkler. Blockages may occur in the body
of the sprinkler and may require removal of the sprinkler to
dislodge such obstructions. Use the special nozzle key supplied
by most manufacturers and avoid the use of monkey wrenches
when removing or assembling nozzles. Take care to reassemble
the nozzle correctly; this applies particularly to spreader nozzles
and to nozzles where the drive is secured by the hammer striking the nozzle.

c) Sprinkler Leakages

Ensure that there is no leakage of water from between the
body and tube of the sprinkler, caused by sand or grit lodging
in the sleeve between the tube and the base. An accumulation of
sand or grit in the sleeve will prevent the sprinkler from turning.
Sand or grit is dislodged by moving the sprinkler body up and
down without closing the riser valve. This flushes the sleeve.
Repeat this procedure until the sprinkler is free to turn unimpe-
ded and the leakage has ceased. These leakages are also caused
by a worn out seal. If so, replace with a new seal at the earliest
opportunity.

d) Misalignment

The sprinkler may cease to turn due to misalignment of the
spoon, the wedge or the hammer. Further, the rubber or plastic
buffer may be worn out. In the event of misalignment, remove
the sprinkler for repair. The buffer may be replaced on the spot.

e) Spring Tension

Ensure that the spring tension is sufficient to return the hammer for striking. If necessary, replace with a new spring. It is

not advised to make field adjustments to the spring tension, unless there is a slotted or castellated spring crown for regulating tension. However, even in such cases the regulating aid is primarily intended for adjusting tension in the event of a change in the nozzle diameter, warranting adjustment of spring tension. The regulating device should be used with caution to adjust the spring in the event of spring fatigue.

f) THE SPRINKLER SHAFT SHOULD BE CHECKED FOR
   SIGNS OF WEARING

Sprinklers should be overhauled before the beginning of the season and checked to ensure their operation within the prescribed pressure range. Nozzle diameters should be spot checked to ensure that abrasive materials in the irrigation water have not caused wearing, enlarging or deforming of the nozzle. Grease or oil should not be applied to the sprinkler, with the exception of giant sprinklers when specifically recommended by the manufacturer.

Care and maintenance of the sprinkler system

a) Couplers are an important part of the system. Treat them with care and give due attention to the pipes.

b) Sealing rings are made of natural rubber and prevent leakage in pipe joints. Inspect them each season for signs of perishing or damage.

c) At the end of a season, remove sealing rings from couplers. To avoid perishing brush in clear water, allow to dry and store in boxes away from light in a cool place.

d) When replacing seals again make sure that the rings seat evenly around the inside of the coupler and that no debris are trapped between the ring and the wall.

Always open and close valves slowly to avoid damage to the pipe system and pumps from water hammer. At the end of the season, check valve stems and seating pads for potting or damage and replace where necessary. When not in use do not seal valves to avoid rubber sticking to the seat.

If sprinklers are moved regularly, the following checks should be made daily:

a) Ensure that nozzles are completely free of obstructions. Do not clear obstruction with a sharp metal object such as a screw driver as this may damage the bore of the nozzle.

b) Ensure that the swing arms are free to rotate and see that the main body of each sprinkler is free to rotate on its main bearing.

At the end of the irrigation season, the following checks should be made before laying the equipment in storage:

a) Check all sprinklers for wear in the main and swing arm bearing; wash and brush them in clean water to remove any grit.

b) Check nozzle bore for damage and wear. Silt and sand particles in irrigation water can cause wear and increase the size of the bore. This will affect the performance of the sprinkler nozzles. Wear can easily be checked using the shank of a twist drill of the same size as the nozzle.

c) Check tension spring on swing arms. If a new one is fitted, set the tension to give the recommended speed or rotation using a spring gauge.

Pumping plant care and maintenance

Installation

a) Avoid high suction heights by locating the pump close to the supply water level, but keep it on firm level ground and not at a risk from flooding.

b) Lay pipe work in a uniform upward slope. Avoid high spots where air can collect. Air collection will reduce the pump efficiency and may cause depriming.

c) Place the inlet of the suction pipe at least 0.6 m below the lowest water level. This stops air from being drawn into the pipe through a whirlpool or vortex. If insufficient water depth is available, use an empty oil drum or barrel sunk into the bed of the stream to provide a sump.

d) Avoid short pipe bends close to the pump. They will disturb the flow of water causing a noisy operation and loss of efficiency. If a bend is necessary keep it well away from the pump.

Ensure that all pump components are lubricated as recommended by the manufacturers. Many pumps have water lubricated seals. These seals must not be lubricated with oil. Do not expect a pump to deliver water at a higher rate or pressure than it is designed for e.g. by adding more sprinklers or pipes. Such over-loading will cause excessive water load on the pump and shorten its working life.

Drain all water from the end of an irrigation pump. This will prevent corrosion. Open the delivery valve for draining. Always leave the delivery valve slightly open when not in use to avoid the rubber seals from sticking to the sealing. At the beginning of a new season always rotate the impeller by hand before start-ing. The pump may have seized while not in use. Therefore re-starting it, without the precautionary measures may damage both the pump and the power unit.

The general principles regarding maintenance of the pipes and fittings and sprinklers are given below:

1) *Pipes and fittings*

Though pipes and fittings require no maintenance they have to receive attention on the following lines:

a) Occasional cleaning of dirt or sand out of the groove in the coupler in which the rubber sealing rings fit. Any accumu-lation of dirt or sand will affect the performance of the rubber sealing ring.

b) Bolts and nuts to be periodically checked and tightened, and not to lay pipes on new damp concrete or on piles of ferti-liser.

2) *Sprinkler heads*

The following attention should be given:

a) When moving the sprinkler lines, make sure that the sprin-klers are not damaged or pushed into the soil. Do not apply oil, grease or any other lubricant which may stop them from working.

b) Check the washers for wear once a season, (and this is espe-cially important when water is sandy).

In general check all equipment at the end of the season and carry out any repairs and adjustments to be made, by ordering spare parts not in stock immediately, so that the equipment is ready for use at the commencement of the next season.

Storage

The following points are to be observed while storing the

sprinkler equipment during off-season:

a) Remove the sprinkler and store in a cool, dry place.

b) Remove the rubber sealing rings from the couplers and fittings and store them in a cool, dark place.

c) The pipes can be stored in the open (not in the proximity of fertilisers) placed in racks, with one end higher than the other.

d) Disconnect the suction and delivery pipe from the pump and pour in a small quantity of medium grade oil at the two ends. Rotate the pump for a few minutes, and then close the suction and delivery ends with stoppers. This will prevent the pump from rusting. Grease the shafts.

e) Protect the electric motor from the ingress of dust; damp and rodents.

The depreciation factors for various sprinkler systems equipment is given in Table 18.

Table 18. Depreciation factors for sprinker equipment

1.	Water supply	
	(a) Well and casing	15 years
	(b) Housing	15 years
	(c) Reservoir—years before silting up	
	(d) Farm ditches—annual maintenance	
2.	Pump, turbine	
	(a) Bowl (estimated 50% of pump cost)	10,000 hrs or 5 year
	(b) Column	20,000 hrs or 10 years
3.	Pump, centrifugal	20,000 hrs or 10 years
4.	Power transmission	
	(a) Gear head	20,000 hrs or 10 years
	(b) V-belts	2,000 hrs or 1 year
	(c) Flat belt, rubber, fabric	4,000 hrs or 2 years
5.	Electric motors	30,000 hrs or 15 years
6.	Diesel engines	16,000 hrs or 8 years
7.	Pipe	
	(a) Aluminium (main and laterals)	10 years
	(b) P.V.C.	4 years
8.	Sprinkler heads	5 years
9.	Misc. fittings aluminium (valves, couplers, elbows, etc.)	10 years
10.	Misc. fittings, PVC.	4 years

Source: Sprinkler Irrigation Guidebook, USAID, Washington, 1962.

CHAPTER 13

Economic Feasibility and Implications

Feasibility

The economic feasibility of any irrigation project requires a study of agricultural production and costs without irrigation, and the expected results with irrigation. India is blessed with good natural resources including waters. Hence it was not a problem in getting water for irrigation. Even now only about 60-70 per cent of the harnessable water has been utilised and action is being taken to utilise all the water. Some States are facing acute water shortage and hence there is a need to introduce, sprinkler and drip irrigation systems there. Since these systems are costly for the poor and small farmers, the farming community was not very familiar with them. The Governments concern of water scarcity has led to incentives to farmers, if they take up sprinkler irrigation. There is a vast scope of introducing these advanced methods of irrigation in this country not only in well irrigated areas, but also in the canal and tank water irrigated areas for close growing crops.

Sprinkler irrigation has been in wide usage in different parts of the world, attracting wide adoption in the northern States of India in the last few years. In the southern region, this system has been in use for a long time for plantation crops in the hills. Due to scarce water conditions prevailing many parts of the country, it is only appropriate to use the sprinkler system in these regions.

The system can be tried not only for groundnut, millets and

117

sugar cane, but all closely spaced crops. Even for paddy in upland condition the sprinkler system has been found to be useful. The systems costs can be reduced by keeping the main pipes permanent and laterals as portable. The cost per ha is nearly Rs. 7500 to 12,500 excluding the cost of motor and pumpset. While adopting sprinkler irrigation for sugar cane the riser pipes have to be fitted at two metres height for covering the entire area, and suitable sprinkler spacing is to be tried.

By using sprinkler irrigation, production increase is claimed to be 40 per cent in some places and for some crops. This seems to be on the higher side and may not be consistently obtained in all locations. Sprinkler irrigation helps to give supplemental irrigation in times of drought, and further helps to distribute the available water efficiently. There is full control and the limited water can be used as and when required. The major advantages in adopting sprinkler irrigation are as follows:

Sprinkler irrigation avoids the possibility of excessive watering leading to the rise in subsoil water level and subsequently to waterlogging. In cases, where the underlying impermeable strata exist near the ground surface, the controlled watering by sprinkler irrigation also checks the rise in the ground water table. This very favourable conditions for agriculture can be introduced through sprinkler irrigation. In short, sprinkler irrigation is considered of importance and of advantage to one or more of the following situations:

1) Where the land is highly undulated with steep gradients which require a high cost of levelling.

2) Where the soils are light and sandy and the rate of percolation is excessive.

3) Where the profile of light soils is shallow with underlying hard and compact impermeable strata which leads to the waterlogging conditions under flow irrigation.

4) Where the available water supply is scarce or inadequate so that the economic use of water becomes imperative.

5) Where by any means the gravity irrigation is not feasible and intensity of irrigation is lower.

There are other advantages of sprinkler irrigation as well:

i) Loss of water (which can be as high as 35 per cent) occurring as seepage in earthen channels during conveyance of water is eliminated (this loss in conveyance can be made to disappear in surface irrigation also by using either lined channels or under ground pipes).

ii) Land with undulating topography can be irrigated with sprinklers with only minor levelling required either for removing excess rainfall water, or for proper functioning of seed drills and other such implements, whereas, for uniform water application, no levelling is required which is a costly affair.

iii) There is a loss of production area, by making bunds and channels. This may be as high as 10 per cent which is lost either in making earthen conveyance channels or irrigation furrows. The whole of this area can instead be brought under cultivation if sprinkler irrigation is practiced.

iv) A sprinkler can apply irrigation at a rate which is less than the infiltration rate of the soil. This helps in the total elimination of runoff losses. There is a precise control over the depth of application, thus eliminating deep percolation losses, which may be more predominant in light soils under surface irrigation. Many times (as while seeds germinate) only light irrigation is required which can be effectively applied with a sprinkler, whereas a surface irrigation system can be designed only for a given depth of application. Shallow irrigation required by a crop in its early stage of growth can be easily applied with a sprinkler, which may be practically impossible with surface irrigation because of the limitations due to steep slopes or porous nature of ground.

v) Fertilisers and other such treatments like soil amendments can be applied in solution along with irrigation water from the sprinklers. These materials are distributed uniformly and can be leached to the desired depths.

vi) Sprinklers give a gentle rain that does not clog or compact the soil. This ensures better and quicker germination of seeds and gives more plants per acre. The quality of some crops such as fruits, vegetables, potatoes and groundnuts can be improved with them. A sprinkler has also been used successfully for protecting crops against frost, when the temperature drops below freezing point. For frost protection water must be sprayed continuously on plants. The application rate from sprinklers has to be kept low as frost may continue for a few days.

vii) In case of surface irrigation the levelling required in an undulating area, disturbs the top fertile soil which is not so in sprinkler irrigation.

viii) Harmful ditch weeds under surface irrigation do not appear with sprinkler irrigation.

ix) In the sugar cane crop, a sprinkler also helps in increasing the sugar content. This system may be adopted with considerable success in porous and sandy soils. In places where the wind velocity is very high the selection of the system should be such that the loss of water while spraying is at the minimum. This can be achieved by resorting to night irrigation and by selection of low pressure nozzles with narrow spacing of the sprinklers, the uniformity of application can be improved.

The increase in production for all the crops depends not only on the method of water application but also on the type of crop, soil, fertilisers etc. Even if we do not take into account the increased production factor, the use of the sprinkler system can be justified because of its better utility in undulating terrains, porous soils, water scarcity area, high mobility, easy operational utility, full control over water application, and high uniform application and water use efficiency. That the yield increases by the use of a sprinkler and is not consistent in all areas is due to the yield being influenced by other factors, such as, sunshine, humidity, climate, texture and fertility of soils in the different tracts.

The water use in the sprinkler and surface methods as found by various experiments in different research centres are given in Table 19.

Table 19. Water use in sprinkler and surface methods

Crop	Water depth required in sprinkler irrigation	Surface (control)	Water saving per cent
Bajra	7.82	17.78	56
Jowar	11.27	25.40	56
Cotton	29.05	40.64	29
Wheat	14.52	33.02	56
Barley	7.82	17.78	56
Gram	7.82	17.78	56
Potato	30.00	60.00	50

Source: Malhotra, A.N. Sprinkler irrigation case study. Proceedings of the Seminar on Sprinkler and Drip Irrigation Systems, Delhi, 1984.

Economics

If 5 ha is cultivated as follows:

In one Year	Approx. net income in rupees
for 2 ha cotton in 2 seasons	2 × 4,000 = 8,000
for 2 ha groundnut in 2 seasons	2 × 3,000 = 6,000
for 1 ha sugar cane	1 × 9,000 = 9,000
	23,000

The approximate net income from the field in one year will be about Rs. 23,000. Since the water saving is about 40 per cent for all crops, the area of cultivation can be increased by 50 per cent more. Hence the income from the land is increased. The benefit cost ratio works out to 1.7 even with a very conservative estimate of income from the produce as given below:

Benefit cost ratio	(Crop) Cotton, groundnut, and sugar cane for the whole area
Water available	5 ha cm
Area (land) available	Unlimited
Well	Existing well can be utilised
Cost of sprinkler Rs. 27000/- Depreciation 20 years	1350
Interest at 12% over 27000	3200
Maintenance	
Sprinkler assembly at 3%	810
Well structure motor etc.	205
Energy charges at 125/HP for 3 Nos	375
Field management charges at 120/ha	600
Total	6740

Assuming that the water saving using sprinkler irrigation is 40 per cent utilising this water for irrigating more area even if no increase in yield is taken into account the benefit of using a sprinkler is as follows:

Total income from field for one year	23,000
Total income from the farm if 1.5 times the area of all crops are cultivated in the same proportion using the same quantum of water	34,500
Net annual benefit	11,500

$$\text{Benefit cost ratio} = \frac{11,500}{6740} = 1.705$$
$$= 1.70$$

Sprinkler irrigation can be adopted even in dry tracts by constructing wells as sources of supply to a community for supplemental irrigation. In Karnataka, the Government has a project of this nature and it is extremely popular.

An increased awareness in the utility value of the sprinkler system is taking place among the agriculturists of India. A recent cost benefit analysis made by Pantnagar University after conducting field trials on maize, with and without the use of sprinklers gives valuable deductions. These observations point to the sprinklers being slightly costlier than the surface irrigation method for normal cropping programmes. However, the quality of the crop with the sprinkler is far superior. The control over precipitation is a marked factor in a farm irrigated by sprinklers. A rational look at the cost benefit analysis must take into account the cost of irrigation water as also the price of quality produce. In certain parts of the country, the irrigation water is metered on a cusec/cumec basis whereas in other States it is charged on an area basis. Therefore, the cost of irrigation water does not get sufficient accent in the latter case.

What is significant is that farms which cannot be cultivated by normal methods of irrigation could be cultivated with sprinklers, given a certain set of conditions. It was found in certain cases, that the cost of levelling a fallow land is much more than the cost of installing a sprinkler system. Such lands could also be brought under cultivation by employing the sprinkler system.

An important technological revolution is quietly overtaking irrigated agriculture in India. This revolution is the change in irrigation technique to sprinklers and drip. The sprinkler irrigated area was very meagre some 10 years ago (not including plantation crops). It is now practiced in about 2.5 lakh ha of land. More farmers are now interested in coming forward to change their traditional method of surface irrigation to sprinkler and drip irrigation, since the water table is going down and the availability of water is dwindling. It is believed that this trend will continue and accelerate in the future. The reasons for the rapid growth of the sprinkler irrigation, and to point out some of its potentially important impacts on issues of agricultural policy, are covered in this chapter.

There are many types of sprinkler systems available, but only the rotating or revolving system is common, and that too the conventional small rotary sprinkler. The boom type self-propelled can be used, but is advantageous for large sized holdings varying from 20 to 60 ha. All the systems may be regulated if necessary by clocks or by a variety of sensing devices detecting soil moisture and field temperature, and by which time these devices materialise, it could be possible to automise irrigation.

There are many areas with light soils, sandy tracts, sloping lands unsuited for irrigation. In some cases, surface irrigation is followed with very low efficiency even in these tracts. If sprinkler agriculture is practiced, there will be a significant change in the entire agricultural system.

INVESTMENTS

The sprinkler irrigation system may cost about Rs. 7500 to Rs. 12,500 per ha. With proper management and with only minor repairs, its expected life may be in the order of 15 to 20 years. This cost is higher than the cost of land levelling and channel formation normally associated with surface irrigation. However, these systems are installed on comparatively low valued lands, so that the total fixed investment, inclusive of land costs is somewhat lower for sprinkler agriculture.

Variable Costs

The light, sandy soils best suited to sprinkler agriculture do not require the intensive soil preparation. The tillage cost for sprinkler agriculture may be only about 50 per cent of the surface irrigated crops. In addition, one must consider the savings in irrigation labour between the two systems. Water use efficiency in a sprinkler is much greater than the surface irrigation, and as such less water is applied. This compensates for the little difference in the net costs of water between the two systems.

Yield

It is difficult to meaningfully compare yields between furrow and sprinkler agricultural systems because plant seed and spacings and fertiliser applications differ as well as irrigation technique. It is estimated that the yields from sprinkler agriculture are consistently over 15 to 20 per cent higher than from surface irrigation, though the figures quoted vary from place to place. There are certain agricultural reasons for increased yields under sprinkler irrigation. Sprinkler irrigation affords better control of wind erosion of the field, it provides an ideal seed bed for the young plant, it may be used to prevent damaging crust formations over young shoots, it may be used for field temperature regulation and other physical reasons may be adduced. But the important factor is that there is a good control of irrigation water application. With the very best prepared field and irrigated by a very good irrigator, surface irrigation will have areas which are over-irrigated and areas which are under-irrigated. In the former, seeds are washed out and plant nutrients carried away resulting in plants simply drowning. In the latter, portions of the field dry out permanently and wilt to an irrevocable degree of damage to the plant. No irrigation system results in absolutely uniform coverage of the field. But the sprinkler system approach is ideal with less managerial input than the conventional surface system. In view of these findings, together with the agricultural advantages above, it is not surprising that yields under sprinkler agriculture are substantially higher.

Labour and Management

For a variety of social and economic reasons which are too complex to go into here, the supply of agricultural labour and the quality of labour available for irrigation is rapidly diminishing. Therefore, an agricultural farm relying on hired labour is interested to go for sprinkler irrigation. By centralising control of irrigation in large farms, by a few qualified people, this technology enables farms to realise significant economies of scale latent in irrigated agriculture.

There are some of the potentially important impacts of modern irrigation techniques on certain problems of agricultural policy. There are water conservation and development, rural development and agricultural development.

WATER CONSERVATION AND DEVELOPMENT

It is a fact that the efficiency of sprinkler irrigation is generally higher than surface irrigation. How much higher whether 30 per cent or 50 per cent is a matter of detail. It depends on soil conditions, climate, the type of surface irrigated practiced. In a most comprehensive review of sprinkler irrigation, Dr. Jack Keller concludes "When good water control was practiced through the use of lined channels or pipes to distribute water to well prepared fields having deep rooted crops growing in medium textured soils, the farm irrigation efficiency differences between different methods were negligible. However, under ordinary field conditions and practices, sprinkler irrigation efficiencies ranged from 25 to 400 per cent greater than those of other methods". In view of the higher irrigation efficiency, less water must be developed per ha in a sprinkler. It means that less water must be distributed through a transportation network to the land. By conveying through pipe, not only the seepage losses of unlined canals but the evaporation losses of lined canals (as high as 10 per cent) are avoided.

Sprinkler irrigation also offers potentially large economics in meeting the drainage and leaching requirements of irrigation projects. In areas where little or no leaching is required, it may not be necessary to provide any drainage facilities at all because of high field irrigation efficiency of the sprinklers. It is also shown that crops can tolerate a more saline condition of irrigation with sprinklers. There is some evidence that sprinklers leach more efficiently than do furrow or flood irrigation. Finally under normal conditions a well designed sprinkler based project could irrigate more land with a given amount of water or irrigate the same land with less water than a project operated by the surface method.

Another point to be considered is that if the real cost of water was charged to the agricultural farm, there would be a wholesale conversion to either sprinkler or to some modern irrigation device in the country. But no one with his eyes on the political realities would expect this eventuality in the near future. Therefore, we must turn our attention to other inducements for water conservation in the agricultural fields. These new irrigation technologies, combined with appropriate institutional modification, could conceivably generate water conservation to eliminate the need for large scale water development.

RURAL DEVELOPMENT

Large areas in the country depend on the vagaries of the monsoon and the yield is very poor. However, there is ground water in these watersheds. By taking this water and introducing sprinkler irrigation, large areas in the rural belt previously unproductive, could be brought under production and income to the communities, as also generate employment opportunities in the rural areas. This is a different social order associated with large agricultural operation.

AGRICULTURAL DEVELOPMENT

At first sight, it may seem paradoxical to think that a capital intensive, labour saving technology may be most efficient in developing an economy short on capital but with more labour availability. Economists have long since ceased to regard the supply of labour as simply the number of live bodies. Instead labour has to be regarded as human capital—a complex of talents, skills and incentives without which simple manpower is virtually useless. The essential advantage of sprinkler agriculture in an under-developed/developing nation is the same as in a developed nation. It divorces the crucial growing period of the crop from the vagaries of human ignorance and sloth. Planting and harvesting can be accomplished through traditional means, but yields are dependent upon irrigation and fertilisation. Given the all important control over these two factors under sprinkler agriculture the yields could be brought to the level of the developed nations.

CHAPTER 14

Research and Development Needs

India is a vast country with varying climate and rainfall. The world's highest rainfall is at Chirapunji and the lowest is in Rajasthan. The soil and the depth of the soil are different in different zones. There are problem soils and eroded lands and these lands are to be reclaimed and brought under cultivation to feed the growing population. Because of the concentration of rain in most parts of the country during a few months of the year, maximum river flow occurs during that period. At other times, the river flow dwindles and many streams dry up altogether. All the rain water could not be stored in view of the topography, location and political constraints and hence a large quantity of water is lost as runoff. Out of a total of 400 MHm of rainfall, only about 100 MHm of water (surface and ground water) could be harnessed to irrigate about 113 Mha. Though the National Commission on Agriculture has reported that the entire harnessable water resources can be brought to use by the year 2025 A.D., the Government of India is planning to complete all the irrigation projects and to use all available water by the end of the century that is by the year 2000 A.D. The total area which could be brought under irrigation by 2000 A.D. is estimated about 113 Mha.

The net cultivated area is about 145 Mha and this may rise to 155 Mha in another 15 to 20 years. The gross cropped area can touch about 200 to 210 Mha at that time. This indicates that the intensity of cropping for irrigated and unirrigated area together is expected to increase from 120 to 135 per cent. The above data indicates that by utilising all the available water, it is possible to bring only about 50 per cent of the gross cultivable area under irrigation and the balance have to depend upon the monsoon after 2000 A.D.

Further the amount of water diverted to drinking, household, cattle and industrial purposes is about 10 per cent at present and this figure will increase to 20-25 per cent in the next 15 to 20 years. On one side more area is to be brought under irrigation to increase the production for the growing population and at the same time, allocation of water is inducted from 90 to 75 per cent for agriculture in the coming years.

The population is growing rapidly and it is expected to touch 1000 M in the year 2000. The food requirement is about 235 MT. It is possible to increase this quantity by bringing more area under irrigation in the coming years. This is possible only by introducing advanced methods of irrigation like sprinkler and drip irrigation. About 40 per cent of the total production is obtained from the dry land which is about 60 per aent of the cultivated area and the balance is from the irrigated land. The productivity and production can be increased to 150 to 300 per cent by promoting irrigation facilities to the crops.

Though the land and water resources are constant, the demand for the cultivated area can be met by increasing the intensity from 120 to 135 or even 200-300 per cent whereas in the case of water the allotment for agriculture would be reducing as the demand for other purposes is increasing.

In most parts of the country water resources are insufficient to meet the irrigation requirements of the cultivated land. Therefore, it becomes necessary to utilise the available water in these areas so as to secure the maximum crop production per unit of water, extending at the same time the benefit of irrigation to as many farmers as is technically and economically feasible. This requires the adoption of the best technique of irrigating the field, minimising the waste of water in storage and conveyance and adoption of a cropping pattern which takes into account the constraint of limited irrigation supplies to produce maximum benefit.

Since water is becoming a scarce commodity and the need is expanding, the concept of supplemental irrigation/protection irrigation should be introduced. In this concept, water is to be given to the crops only when it is needed to the crop once or twice during its growth and not regularly as done in the case of canal, tank or well irrigated fields. By this, the yield in the entire cultivated area in the country can be enhanced tremendously. This can be carried out especially in a watershed where some ground water is still available. Instead of using the water for intensive irrigation, extensive irrigation as mentioned above can be followed. This could be possible only through sprinkler and drip irrigations.

Whenever an irrigation project is constructed, the tendency of the farmer is to go for paddy with flooding methods where the overall efficiency of irrigation water is only about 30 to 40 per cent. Irrigation methods have to be changed depending upon the crop. Water should be delivered in a controlled manner. Sprinkler irrigation can be introduced for close growing crops like sugar cane, groundnut, cotton, millets, pulses etc. The saving may be about 40 to 50 per cent and hence more area can be brought under irrigation from the same quantity of water.

Research need

There are a number of problems in connection with sprinkler irrigation which should be made the subject of research. Some of these problems have been studied in a limited way but the information so far secured is far from complete. The Central Water Commission and the Central Board of Irrigation and Power are interested to get detailed information about the comparative performance, water requirements between surface and sprinkler irrigation. Many agricultural universities and engineering universities are taking interest in the last few years and more information will be generated in order to adopt this system in the coming years.

The following are the areas in which research informations are required in the sprinkler system:

1) EVAPORATION LOSSES
   a) Direct loss between sprinkler head/soil/plant.
   b) Indirect loss from vegetation:
      i) During the period of application.
      ii) After the application is completed.
   c) Indirect loss from the soil surface.

2) EFFECT ON CROPS

a) Under what condition, if any is pollination interfered with?

b) Under what condition the yields are increased as compared to those under surface irrigation and why?

c) What effect does sprinkler have on plant disease, insect etc.?

d) What effect does a sprinkler have on ripening and quality of the plant?

3) EFFECT ON SOILS

a) What are the effects of rate of application?

  i) By better moisture conditions.

  ii) By reduced root pruning.

4) EQUIPMENT

a) To what extent and how can distribution pattern of an individual and groups of sprinkler be improved?

b) What are the possibilities of large sprinkler and mini-sprinklers?

c) What life can be expected for different type of equipment under different conditions?

d) Under what conditions can fertiliser, insecticides, fungicides be successfully applied by irrigation by the sprinkler system?

5) USE

a) What irrigation efficiency can be expected with sprinklers under different conditions?

b) Can over-all savings in cost be achieved, by changes in design perhaps by higher pressures and wide spacings?

6) WATER CONSUMPTION

a) Water requirements of various crops by sprinkler, perfro-spray, mini-sprinklers.

b) Advantages of sprinkler irrigation for various crops, compared to surface irrigation.

7) COMPARATIVE ECONOMY

It is therefore necessary to expand the research programme in the use of sprinkler irrigation. The basic research, the effect of sprinkling on crops and soils and water efficiency studies should probably be undertaken by the agricultural university and other institutions. The manufacturers should continue to develop new and better equipment. Finally the users and farmers will determine its overall economy and usefulness.

DEVELOPMENTAL NEED

The development/irrigated area under sprinkler is very meagre. It is estimated that only 0.25 Mha is under sprinkler in the country whereas the irrigated area is about 70 Mha. In Israel about 90 per cent of the area is covered by the sprinkler system and in the U.S.A. it is about 35-40 per cent. Indigenous manufacture of the system started about 20 years ago, and it has all the facilities with the product updated. But there are only a limited number of firms that are involved in this field since the demand is not commensurate with the need.

Sprinkler irrigation is widely used in plantation crops like, coffee, tea, rubber, cardamom etc. The crops are grown in high rainfall areas in the hills. The rain occurs for about 7 to 9 months, but during January to May there may not be enough moisture for the crop. Further, crops like coffee require 'showers' during this period for its flowering and for good yield. For cardamom, if one or two irrigations are given during this hot season, the yield of the crop can be increased enormously. In addition, the development boards like coffee, tea, and cardamom are having schemes for supplying the sprinkler equipment on a hire-purchase basis and also subsidising about 25 to 50 per cent of the total cost to the farmers. Commercial banks are also coming forward to finance the viable schemes for introducing sprinklers on a large scale. This is in a way helping the farmers to go in for this system.

Due to scarcity of water the State Government to Haryana is launching big schemes to provide sprinkler irrigation to the farmers. To give incentives, the Government is subsidising about 25 per cent of the cost to the farmers and the balance is given on loan by the banks. About 10,000 units have already been installed in this State. Similarly the Government of Gujarat Maharashtra, and Karnataka are taking steps to popularise this system on a large scale. The Government of India is also encouraging the State Government to introduce this water saving method by providing 50 per cent of the subsidy to the State Governments.

The economic and other advantages of the system are favourable. Sprinkler irrigation can be introduced on a large scale throughout the country for closely spaced crops. This will facilitate the increase of the area of irrigation at least by 50 per cent.

Large scale demonstrations for various crops can be arranged in different locations with sprinkler irrigation. A programme may be worked out to bring at least 2 to 3 per cent of the irrigated area by the year 2000. This will not only help to increase production but also increase employment opportunities both in rural and urban areas.

CHAPTER 15

When to Irrigate and How Much

When to irrigate and how much water to apply are the two most important questions to be answered when irrigation is given to any crops whether it is from surface or from overhead or from drip irrigation. This chapter outlines the guidelines and principles to answer these questions. Combined with this, the knowledge, experience and judgment of the farmer count in applying inputs for satisfactorily achieving the irrigation water management.

SOIL

The soil is a store house of plant nutrients, an anchorage for plants and reservoir that holds the water needed for plant growth. The amount of water which any soil can hold for plant use, is determined by the soil's physical properties. The soil texture is the most influential factor in determining the available water holding capacity (AWC) of the soil. Many soils will have varying textures at different depths. Table 20 gives the general range of AWC per cm/30 cm of depth for various soil textures. Soil textures of your field can be ascertained from the local agricultural extension officers or from specialists in an agricultural university.

ROOTING DEPTH

The depth of soil from which a crop can extract water is the effective crop rooting depth. This depth will vary with the stage

133

Table 20. Available water holding capacity for various soil groups

Soil classification	Available moisture	
	Range cm/30 cm	Average cm/30 cm
Very coarse to coarse textured sand	1.25-2.6	2.25
Moderately coarse, textured sandy loams and fine sandy loam	3.10-4.50	3.75
Medium texture—very fine sandy loam and silty clay loam	3.75-5.75	4.75
Fine and very fine textures—silky clay to clay	4.00-6.25	5.25
Peats and mucks	5.00-7.50	6.25

of crop growth. Rooting depths in some cases may be restricted to a depth considerably less than the normal root depths by soil limitations. like hard pan and high water table. Normal rooting depths for several matured irrigated crops, growth on deep, permeable, well drained soils are given in Table 21.

TOTAL AVAILABLE MOISTURE

The total amount of water available for plant use in the root zone is the sum of the available water holding capacity per metre for the various soil types within the effective rooting depth. To give an example, the total available moisture in the root zone, by assuming the crop cotton with a 60 cm rooting depth in a soil is calculated with the following characteristics:

Soil	Depth cm	Layer thickness in cm × AWC per metre	Available moisture cm
Sandy loam	0-20	$\frac{20}{100} \times 12.40$	2.50
Silty clay loam	20-50	$\frac{30}{100} \times 16.5$	5.00
Loamy sand	50-60	$\frac{10}{100} \times 8.25$	0.90
		Total available moisture	8.40

Table 21. Normal crop rooting depths

S. No.	Crop	Root depth, cm
1.	Cotton	60-75
2.	Sugar cane	45-60
3.	Groundnut	30-45
4.	Millets/*sorghum*	30-45
5.	*Bajra*	30-45
6.	Banana	45-60
7.	Corn/maize	30-45
8.	Grapes	30-45
9.	Onion	30
10.	Potato	30
11.	Beans	30
12.	Tomato	30-45
13.	Brinjal	30-45
14.	*Bendai*	30

ALLOWABLE MOISTURE DEPLETION

There is an economic optimum amount or level of soil moisture that should be depleted before irrigating. This point is often
described as a portion or per cent of the total available moisture in the root zone. There are many factors that should be
considered in determining this allowable depletion level, the most
important of which is the crop itself. Other factors such as consumptive use rate, available water supply, salinity, drainage and
labour should also be considered. Moisture depletion allowances
usually vary from 25 to 80 per cent. The depletion level for most
of the crop is about 50 per cent but for banana and sugar cane
it is about 30-40 per cent, groundnut 60 per cent and cotton 75
per cent.

WHEN TO IRRIGATE

Irrigation should be scheduled when the amount of water depleted from the soil in the root zone approaches the allowable
depletion level selected. One of the methods in estimating the
amount of water depleted from the soil is the feel method using
Table 14. The depletion level estimates can be made by taking soil
samples with an auger, soil tube or shovel at about 30 cm incre-
ments of the entire crop root zone. This can be quickly obtained
by a neutron probe or by using a tensiometer, gypsum block.

The total amount of moisture depleted is the sum of the entire crop root depth. This is also the net amount to be replaced by irrigation in order to refill the soil profile to its field capacity. Additional allowances must be made for deep percolation, leaching etc.

Example: Estimating when to irrigate

Assume the following:

The same soil and crop as in previous example. An allowable moisture depletion of 60 per cent of the total available in the 60 cm root/zone.

Allowable depletion level: $0.6 \times 8.40 = 5.04$ cm. This means that irrigation should begin when 5.04 cm of moisture has been lost by this calculation.

Leaching requirements

All irrigation water and soils contain some soluble salts. Plants use the water and most of the salts remain in the soil. If not removed periodically, the salts will accumulate to a point that is toxic to the plant.

Application of additional water will flush out the salts from below the crop root zone. This can be done as part of the normal irrigation application or as a special programme. The amount of water required to obtain a satisfactory salt balance in the crop root zone is called the leaching requirements and depends on the amount of salt present in the soil, the irrigation water quality and the crop grown. During drought years, leaching of salt is postponed without any problem.

Irrigation efficiencies

One of the more important factors affecting the water requirements for any irrigation system is the irrigation efficiency. Water application efficiency is the ratio of water stored in the soil root zone and utilised by the crop to the water delivered to the field and is the commonly used measure of irrigation practice. However, application efficiencies may be high and irrigation practice poor if the water applied is not uniformly distributed throughout the field and the root zone of the soil. For this reason application efficiency and the distribution efficiency are treated separately and then combined to give an overall sprinkler irrigation efficiency.

Application efficiency

Application efficiency in the sprinkler irrigation is the ratio of the amount of water reaching the ground surface as measured by the sampling cans to the amount being discharged from the sprinkler nozzles. If there is no loss of water by surface runoff and deep percolation, application efficiency as thus defined gives an indirect measurement of water losses by wind drift and evaporation. While these losses do not represent a great percentage, they are important to consider and should be held to a minimum. Since costs of applying water by sprinkling are relatively high, any waste represents an economic loss.

Evaporation from sprinkler spray while still in the air was calculated by Christiansen to be about 2 per cent provided the spray is not broken up into a fine mist which may drift away. The greater portion of the evaporation loss will be from wetted foliage and soil surface since the exposed surfaces are extreme and water will continue to evaporate for some time after irrigation. However, this evaporation tends to decrease transpiration and may be partially effective in meeting the water needs of the crops.

Among the climatic factors which will affect the efficiency of application are relative humidity, rate of application and temperature. All these are known to influence evaporation. Sprinkler spacing, operating pressure and air movement will also affect application efficiencies, but may have a more pronounced effect on the distribution pattern.

Distribution efficiency

The aim of good sprinkler irrigation is to prevent parts of the field from being under irrigated while other parts are over-irrigated. Lack of uniformity can result in areas of poor vegetative cover and low production. Distribution efficiency as the term implies gives a measure of the ability of a system to apply equal amounts of water to all parts of the area covered. Uniformity of water application by sprinkling is affected by pressure at the nozzle, spacing of the sprinklers and wind movement.

Overall efficiency

In determining system capacity and depths of water to apply, both application efficiency and distribution efficiency should be considered. The product of the two efficiencies gives an overall system efficiency which would ensure that not only sufficient quantities are applied to meet crop requirements, but that all parts of the field receive adequate water.

How much to apply

The gross amount of water to be applied is the net amount required to refill the crop root zone plus the amount required for leaching divided by the overall efficiency.

$$\text{Gross application required} = \frac{\text{Depleted moisture to be replaced} + \text{leaching requirement}}{\text{Overall efficiency}}$$

The crop under irrigation may suffer from lack of water due to any one of the following reasons:

a) starting too late at the beginning of the irrigation seasons;

b) too many days between irrigation;

c) applying less water than the crop uses;

d) failing to irrigate the soil profile completely at any irrigation; and

e) poor distribution of water from the sprinkler.

We will save labour, time, water and pumping costs if we allow long intervals between each irrigation, but we will not necessarily get the highest yields. As a general rule, use up to one-half of the available moisture at one-third root zone depth in the crop row before irrigating again. Some small fruits and vegetables require irrigation at wetter moisture levels.

Enough water should be applied at each irrigation to bring the full depth of the ultimate rooting zone of the crop to field capacity. Once the soil has been wetted to field capacity the ideal amount of water to apply at each irrigation is that amount used by the crop since the last irrigation, plus that which is lost through evaporation and distribution during irrigation. Applying less water will leave dry areas below the plant root resulting in shallow root systems. Plant roots will not grow into dry soils even to reach the moist area below.

Light, frequent irrigations are ordinarily necessary on shallow or sandy soils. Although not usually necessary on heavy soils, they may not be harmful, unless certain diseases caused by higher humidity are present. Since most systems are designed to apply the proper amount of water during peak crop use, either the time between irrigation may be lengthened or the amount of water applied per irrigation may be reduced during the rest of the season.

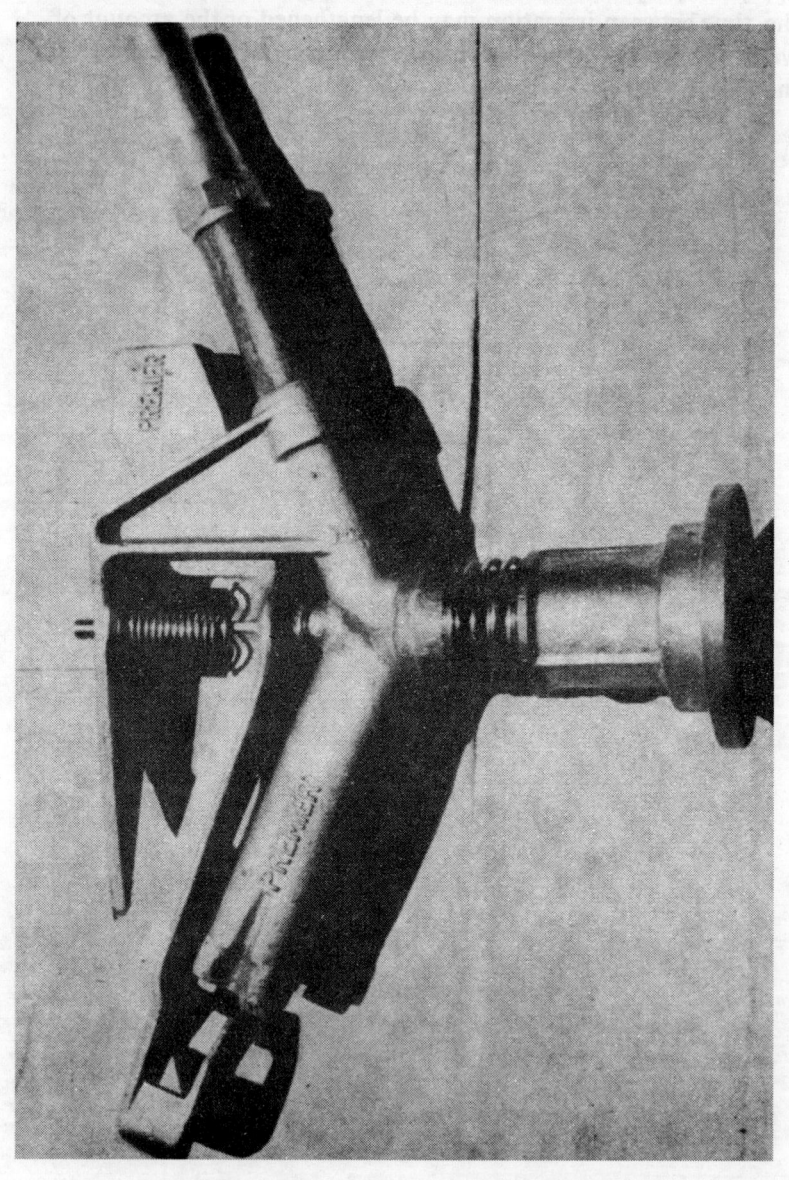

Model 400 Premier sprinkler (sprinkler gun)

Sprinkler irrigation for forage crops

Sprinkler irrigation for vegetables

Sprinklers operating on groundnut crop

References

1 American Embassy. *Water resources development in India—Problems and Prospects*, June 1980, New Delhi.

2 California Department of Water Resources. *Irrigation Where and How Much?*

3 Colorado State University, Tech. Bulletin 128. *Evaluating Water Distribution of Sprinkler Irrigation Systems*, 1976.

4 Duffin, R.B. *Planning Sprinkler Irrigation*, OSU Extension Facts.

5 Elhanani, S. *Sprinkler Irrigation*, Ministry of Agriculture, Tel Aviv, Israel, 1961.

6 Hague, R.M. *Irrigation of Agricultural Lands*, No. 11 in the series—Agronomy American Society of Agronomy, Madison, 1967.

7 Halevy initials and Boaz M. *Sprinkler Ten Lines for Row Crops Irrigation*, Ministry of Agriculture, Tel Aviv, 1973.

8 Hurd Clarence J. *Sprinkler Irrigation*, Guide Book, USAID, Washington, D.C. 1969.

9 IARI. *Training Notes on Sprinkler and Drip Irrigation*, 1985.

10 Jain, B.K.S. *Sprinkler Irrigation Techniques*, Bombay, 1961.

11 Malhotra, A.N. Sprinkler irrigation case study. *Proceedings of the Seminar on Sprinkler and Drift Irrigation Systems*, New Delhi, 1984.

12 Melvyn Kay. *Sprinkler Irrigation Equipment and Practice*, English Language Book Society, London, 1984.

13 Michael, Shrimohan and Swaminathan. *Design and Evaluation of Irrigation Methods*, IARI, New Delhi, 1972.

14 Michael, A.M. *Irrigation Theory and Practices*, Vikas Publishing House Pvt. Ltd., New Delhi, 1978.

15 Ministry of Irrigation, Delhi. *Proceedings of the Seminar on Sprinkler and Drip Irrigation systems*, March 1984.

16 Ministry of Irrigation and Power, Govt. of India. *Report of the Irrigation Commission*, Vol. I, New Delhi, 1972.

17 Ministry of Agriculture and Irrigation, Govt. of India, Report of the National Commission of Agriculture, Part V, Resource development, New Delhi, 1976.

18 Rainbird. *Irrigation Equipment*, California, USA, 1977-78.

19 Report of the Pugalur River Pumping Scheme, Karur, 1973 Cooperation Department, Madras.

20 Scot, V.H. *Sprinkler Irrigation*, Circular 456, University of California, 1956.

21 SCS National Engineering Hand Book, Section 15—Irrigation, Chapter II, Sprinkler Irrigation, USDA, 1960.

21a Ibid. Section 85, Irrigation, Chapter 8, Irrigation Pumping Plants, USDA, 1960.

22 Sivanappan, R.K. Sprinkler and Drift Irrigation. *Proceedings of the Seminar on Sprinkler and Drift Irrigation System*, Ministry of Irrigation, Govt. of India, 1984, New Delhi.

23 Sivanappan, R.K. and K.R. Karai Gowder. *Irrigation and Drainage*, Popular Book Depot, Madras, 1977.

24 University of California, Agricultural Extension Service. *Should I Use Sprinkler for Irrigating Vegetable Crops?*

25 U.P. Agricultural University. *Comparative Performance of Sprinkler and Surface Irrigation*, 1970.

26 USDA. *Sprinkler Irrigation*, Leaflet No. 476, 1966.

27 USDA. *Evaluation of Sprinkler Irrigation Systems in Northern Utah*, Bulletin 287, 1954.

28 USDA Bureau of Reclamation. *Sprinkler Irrigation*, 1943.

29 Water and Power Consultancy Services (India) Ltd. Short Term Courses on Sprinkler and Drift Irrigation, New Delhi, March 1985.

Annexure

AVOIRDUPOIS TO METRIC AND METRIC TO AVOIRDUPOIS CONVERSION FACTORS

1) Length

Multiply	By	To obtain
Metres (m)	3.281	Feet (ft)
Centimetres (cm)	.3937	Inches (in)
Millimetres (mm)	.0394	Inches (in)
Feet (ft)	.305	Metres (m)
Inches (in)	2.54	Centimetres (cm)
Inches (in)	25.4	Millimetres (mm)
Miles (mi)	1.609	Kilometres (km)
Kilometres (km)	.621	Miles (mi)

2) Pressure

Pounds per square inch (psi)	2.31	Feet of water (62° F)
	.703	Metres of water (62° F)
	.0703	Kilogramme per sq cm (kg/cm)
	.068	Atmospheres (Atmos)
	2.036	Inches mercury (32° F)
Feet of water at 62° F	.433	Pounds per sq in (PSI)
	.0305	kg per sq in
	.029	Atmospheres
Metres of water at 62° F	1.422	Pounds per sq in
	.10	kg per sq cm
	.097	Atmospheres
Atmospheres	14.7	Pounds per sq in

142

Multiply	*By*	*To obtain*
Kilogrammes per sq cm	14.22	Pounds per sq in
	10.00	Metres of water

3) **Volume**

Cubic feet	7.48	U.S. gallons (U.S. Gal)
	6.23	Imperial gallons (Imp. Gal)
	28.3	Litres (LT)
	.0283	Cubic metre (m³)
U.S. gallons	.1337	Cubic feet
	.833	Imperial gallons
	3.785	Litres
	.0038	Cubic metres (m³)
Cubic metres	35.315	Cubic feet
	264.2	U.S. gallons
	1000	Litres
U.S. quarts	.9463	Litres
Acre feet	1233.49	Cubic metres
	43,560	Cubic feet
	325,851	Gallons

4) **Units of flow**

Cubic feet per second	448.8	U.S. gallons per minute
	373.7	Imperial gallons per minute
	28.32	Litres per second
U.S. gallons per minute	.00223	Cubic feet per second
	.0631	Litres per second
	.833	Imperial gallons per minute
Imperial gallons per minute	1.2	Gallons per minute
	.0758	Litres per second
Litres per second	.0353	Cubic feet per second
	15.850	U.S. gallons per minute
	13.199	Imperial gallons per minute
Cubic metres per hour	.0098	Cubic feet per second

Multiply	*By*	*To obtain*
	4.403	U.S. gallons per minute
	3.666	Imperial gallons per minute

5) Area

Acres	.4047	Hectares (ha)
	43.560	Square feet
	4047	Square metres
	.00156	Square miles
Hectares	2.471	Acres
	107,639	Square feet
	10000	Square metres
	.00386	Square miles

6) Temperature

$$F^\circ = 9/5(C^\circ) + 32$$
$$C^\circ = 5/9(F^\circ) - 32$$

Index